PAST AND PRESENT VEGETATION OF THE FAR NORTHWEST OF CANADA

The far northwest corner of Canada is remarkable for its great diversity of landscapes, flora, and fauna and particularly for its unusual history during the past several millennia. After describing the rich physical and biological setting of the North Yukon and adjacent Mackenzie Delta region, Ritchie provides a detailed account of the dramatic changes in vegetation cover that accompanied the major shifts in climate from glacial to non-glacial regimes. Because a large part of this area remained unglaciated during the latest ice age, a long and rich record of pollen and other plant remains has made it possible to trace the development of tundra and boreal vegetation and to relate these changes to the record of past faunas. Unsolved problems are confronted and the directions for future research suggested.

Written for scholars and students of Quaternary science in botany, palaeontology, climatology, geology, archaeology, and pedology, this book makes a significant contribution to our understanding of the past and present environments of the Canadian North.

J.C. RITCHIE is Professor of Botany at the University of Toronto.

J.C. RITCHIE

Past and Present Vegetation
of the Far Northwest of Canada

University of Toronto Press
Toronto Buffalo London

© University of Toronto Press 1984
Toronto Buffalo London
Printed in Canada
Reprinted in 2018
ISBN 0-8020-2523-4
ISBN 978-1-4875-8130-5 (paper)

Canadian Cataloguing in Publication Data

Ritchie, J.C. (James Cunningham), 1929–
    Past and present vegetation of the far northwest
    of Canada

    Bibliography: p.
    Includes index.
    ISBN 0-8020-2523-4

    1. Paleobotany – Yukon Territory – Recent.
    2. Paleobotany – Northwest Territories – Mackenzie
    River – Delta – Recent. 3. Botany – Yukon Territory.
    4. Botany – Northwest Territories – Mackenzie River –
    Delta. I. Title.

QE931.3.R57 1984   561'.197191   C83-099229-4

To the memory of my father

Contents

Preface

The main incentive to compile this account of the Quaternary history of the North Yukon and adjacent Mackenzie Delta region came from the realization that the accelerated pace of palaeoecological research in northwest Canada during the past 15 years has produced a large body of information and ideas that might appropriately be collated, reviewed, and interpreted; it seems time, too, to outline possible new problems and new directions for future investigation. The first systematic research in the plant palaeoecology of the northwest was undertaken by Dr Jaan Terasmae in the 1960s, and my own small group began work in the Mackenzie Delta region in 1967. But the pace of investigation and number of investigators have increased dramatically since 1970. First, there was some minor palaeoecological activity in connection with the environmental assessments and analyses that accompanied the enquiries into the desirability of building a natural gas pipeline from the northwest down the Mackenzie River corridor to southern markets. These largely uncoordinated, often necessarily hurried investigations yielded a rich harvest of surveys and inventories on many aspects of the physical and biological environment that would probably still not have been started had the question of resource development not arisen.

But the major impetus came in the early 1970s, triggered by the discovery by Dr C.R. Harington (Irving and Harington 1973) of a caribou tibia with modifications to its surface ascribed to human artifice, radiocarbon-dated at 27,000 years before the present. This fossil, discovered in sediments along the banks of the Old Crow River in North Yukon, and Harington's earlier amassing of several thousand fossil bones of vertebrates, some extinct, others apparently modified by early humans, established this northwest corner of Canada as a locus for burgeoning interest by historical scientists. The dual allure, as the avenue for the 'peopling of America' and as the theatre of spectacular extinctions of large animals if not entire biomes, both inflamed the imaginations of scholars and loosened the purse-strings of granting agencies.

In 1974 two large team projects were conceived simultaneously and independently. Both groups spent the field seasons from 1975 to 1980 at work in the North x Preface

Yukon, centred on the confluence of the Old Crow and Porcupine rivers; in tent camps of various sizes were several dozen specialists at all levels in the academic hierarchy. The main projects – the Yukon Refugium Project directed by Dr Richard E. Morlan and primarily under the sponsorship of the Canadian federal government, and the Northern Yukon Research Programme organized by W.N. Irving and J. Cinq-Mars supported through grants to the University of Toronto – completed their field-work in 1980, though small offshoot projects are still active. Those of us in disciplines peripheral to the mainstreams of palaeontology and archaeology happily embraced the remarkable opportunity to take part in this massive research endeavour. It is much too early to know whether the results have merited either the scientific effort or the funding.

Involvement in the University of Toronto project enabled my colleagues and I to push forward the investigations of late Quaternary vegetation history we had begun earlier in the Mackenzie Delta region. Now a reasonably coherent body of botanical information is at hand. In many respects this book is a status report on 15 years of research into the historical plant ecology of northwest Canada. Some general patterns of change have emerged to justify a few conclusions or at least the formulation of testable hypotheses but, equally, unsolved problems can perhaps be focused more sharply if we pause, assess the state of our knowledge, and harness new methods and concepts for the next stage of enquiry.

Final revision of the scientific aspects of the manuscript was done in February 1983, which also served as the latest date for references cited in the text.

My field and laboratory research in the northwest has been supported generously for over 15 years by the National Science and Engineering Research Council of Canada, the Polar Continental Shelf Project of the Canada Department of Energy, Mines and Resources, the Radiocarbon Laboratory of the Geological Survey of Canada, and the Northern Laboratory (Inuvik) of the Department of Indian and Northern Affairs, Canada. The investigations in the North Yukon were made possible by an invitation from Dr W.N. Irving to join the North Yukon Research Programme of the University of Toronto, funded by the Social Sciences and Humanities Research Council of Canada and the Donner Foundation of Canada. My investigations have been supported in several important ways by Scarborough College, University of Toronto, not least of which through the excellent services of the Academic Workshops and the Graphics Department. I have benefited from many discussions and helpful suggestions, from Jacques Cinq-Mars, Owen Hughes, John Matthews, Richard Morlan, Don Norris, Charles Schweger, and John Westgate. In addition to providing congenial, stimulating friendship in field and laboratory, Les Cwynar, Glen MacDonald, Lynn Ovenden, and Ray Spear prepared rigorous-critiques of the first draft of this book. My research assistant, Kathleen Hadden, completed a wide range of tasks, from field coring to pollen counting, tirelessly and meticulously, and Mrs Danielle Carbone typed drafts of the text with patience and skill. Most of all, June

Hope provided both the spark that initiated the writing process and the unflagging support and constancy that saw it through to fruition.

The aerial photographs on pages 212–13, ©1944, Her Majesty the Queen in Right of Canada, are reproduced from the collection of the National Air Photo Library with permission of Energy, Mines and Resources Canada. The book has been published with the assistance of grants from the Natural Sciences and Engineering Research Council of Canada and from the Publications Fund of the University of Toronto Press.

PAST AND PRESENT VEGETATION OF
THE FAR NORTHWEST OF CANADA

1
Introduction

The remote northwest corner of Canada is intensely interesting for several reasons. It has a surprising and impressive array of landscapes, set on a vast scale: the immense Mackenzie River Delta, several remote, scarcely explored mountain ranges, rivers of every size and gradient, and a wealth of rock types and formations, varied soils, and intricate patterns of frost-related surface features.

Climatically it displays some of the steepest gradients and sharpest contrasts in North America, from the continental subarctic regime of the interior valleys in the southwest with relatively snowy winters and long, warm summers, to the harsh arctic conditions of the distal Tuktoyaktuk Peninsula with sparse snow in winter and short, cold summers. Two major ecological complexes are associated with these contrasting climates: the boreal forest and the tundra. Each has a distinctive diversity of flora and fauna, determined by complex interactions of environmental factors.

But the most intriguing feature of the northwest is that it is the only significant part of Canada that has remained free of glacier ice for the past half million years or more. It also formed, at times of lowered sea level, the eastern part of a land mass – Beringia – that bridged North America and Eurasia. It has long been thought that this episodic link between the continents was the point of entry of a significant part of the modern floras and faunas of North America, and probably of early humans. Thus it offers a tantalizing, long history of change in environment, vegetation, fauna, and human presence.

This book deals with aspects of all these topics, but its main thrust is the history of the vegetation. The primary data come from analyses of pollen grains and plant macrofossils in sediments. We will reach back in time to the mid-Tertiary to sketch in the longer historical background of the plant cover of northwest North America, but the temporal focus is the late Quaternary. Not unexpectedly, we will find that the closer we get to the present day the more detailed and informative is the record.

Historical plant ecology, or plant palaeoecology as it is also termed, strives to decipher and interpret 'experiments of history' (Deevey 1969) in an effort to understand the processes and patterns of plant species and community responses to environmental change. After the slow, very gradual changes in climate and tectonic adjustments of the Tertiary, northern regions were plunged into the violently oscillating glacial and non-glacial climates of the Quaternary period, and a vast but often incomplete inventory of physical and biological records was left in the sediments. They are the results of the 'experiments of history' and they can provide the raw materials for elucidating the tempo and mode of plant community interactions, successions, resiliency in the face of instability, and other processes. The task of finding the evidence is not easy because often the containing sediments have been disturbed, redeposited, or removed entirely. And even when the record is at hand its interpretation is often fraught with uncertainty.

None the less, investigation of vegetation history on the time scale of the Quaternary is of central importance not only to plant ecology and phytogeography, but to other historical sciences as well.

Vegetation is the structural and functional basis of terrestrial ecosystems; in addition it is largely from information on past vegetation that our understanding of changes in terrestrial environments unfolds. Terrestrial palaeoclimates are worked out, largely but of course not exclusively, from evidence of former plant communities or from the record of alterations in the ranges of particular species. The historical ecology of animal and human populations draws heavily on palaeobotanical data for developing an understanding of possible causes of past changes in faunas and in cultural development.

In parallel with the greatly increased flow of facts and ideas about the historical ecology of Beringia, plant palaeoecologists find themselves at an interesting juncture in the development of their discipline. Just as the larger, parent science of ecology is entering the post-descriptive phase of its development (Harper 1982), when the preoccupation with distributional and classificatory data sets is giving way to a more precise, experimentally based search for correlations and causes that will lead to useful generalizations, so Quaternary plant ecology is throwing off its obsession with descriptive pollen stratigraphy and addressing more basic ecological questions such as the nature of responses of plant populations to climatic and other perturbations.

New tools and discoveries have contributed to this change, although it is also true that the necessary biostratigraphic phase of investigation is virtually complete in many regions of the Northern Hemisphere, so that the continued accumulation of standard percentage pollen diagrams becomes of questionable value in many areas. One of the new botanical tools that has advanced an understanding of vegetation history is in the pollen influx method introduced by Davis (1967), itself made possible by the application of radiocarbon dating methods to sediment samples. We will discover how pollen influx data have freed Beringian palaeoecology from the deceptive constructs of percentage pollen diagrams, a condition aptly described by Deevey (1969): 'the quantitative nature of pollen evidence is illusory if one does not know from what area or from what distance or from how many plants the fossil pollen came. Given these major uncertainties, the more one works within closed statistical universes of 100% of pollen grains, the greater the risk of self-deception and circular arguments.'

Botanists were intrigued by the ecological and historical problems of Beringia long before palaeontologists and archaeologists sensed its potential significance. One of the leaders of arctic botany, the late Professor Dr Eric Hultén, set out the basic phytogeographical hypotheses suggested by the vascular plant distributional data over fifty years ago and they have remained widely accepted but untested to the present day.

Hultén suggested that Beringia was a glacial refugium for arctic and boreal plants and that early in the present interglacial period they radiated from the several Beringian centres to take up their varied modern arctic, boreal, and arctic-montane distributional patterns. Subsequent floristic research, including much of his own, has fleshed out Hultén's hypotheses, and they are accepted in one form or another by most authorities (Johnson and Packer 1967; Löve and Löve 1974; Murray 1981; Yurtsev 1972; among others). However, this consensus rests more on 'long held belief' (Murray 1981, p 12) than on adequately tested hypothesis, of necessity. So far the 'factual basis of phytogeography' (Godwin 1975) has not been very helpful in confirming or refuting ideas of Beringian plant geography. We will give some brief attention to these questions later in this treatise.

These questions of hypothesis and 'belief' raise an important matter, and before we proceed further we should set out concisely the working principles, assumptions, and limitations of palaeoecology. Full discussion of these topics with particular reference to Quaternary botany can be found in recent contributions by Birks and Birks (1980, Chapter 1), Birks (1981), and Rymer (1978).

Palaeoecology strives to interpret or explain historical evidence in as direct and simple terms as possible – according to the so-called law of parsimony, or Occam's Razor. The explanation should be couched as multiple hypotheses, refutable by independent data sets, and the framework within which the hypotheses are devised is established from the state of knowledge of modern organism-environment relationships and processes. In other words the concept of methodological uniformitarianism, as clarified recently by Gould (1975) and in its particular application to pollen analysis by Rymer (1978), describes our approach whereby we assume that the fundamental principles and laws of natural science have remained constant in time. An example will illustrate the procedure. Several of the data sets to be reviewed in Chapter 5 show changes during the Holocene in the relative frequency and influx of spruce pollen, with associated occurrences of spruce macrofossils. This evidence for extensions and contractions of the range of spruce can be explained by at least three alternative hypotheses: (1) We can assume that spruce has always been in equilibrium with the climate, and knowing from observation and experiment something of the present-day climatic conditions limiting spruce at treeline we can suggest certain climatic changes as determinants of the alterations in spruce range. (2) We can assume that climatic change is not an adequate explanation for the changes but that migration lags due to dispersal rates have occurred and either these or such other biological factors as competition with other species are sufficient to explain the data. (3) Or we can test the idea that both biological and climatic determinants have operated. The hypotheses are tested by examining the chronology of increases and decreases in spruce populations against time-transgressive or synchronous propositions. Hypotheses of climate control can be tested by examining independent data sets from the same area over the same time period (e.g. other biostratigraphic data and / or evidence from physical systems such as sediment composition and structure). This method of reconstructing past environment, referred to as the indicator species approach (Iversen 1964; Birks 1981), uses data on the modern environmental tolerances of species to derive past environments. Its effectiveness depends on the validity of the assumptions that species are and have been in equilibrium with climate, and that there is a direct relationship between one or several climatic factors and species distribution.

This book is an invitation to students of Quaternary science – coming from botany, palaeontology, climatology, geology, archaeology, pedology, and other disciplines – to examine the facts of the vegetation history of this fascinating part of the world; to consider the plant community reconstructions I have proposed and no doubt to conceive of others; to examine and supplement the hypotheses brought forward to explain the facts or constructs; and to weigh the evidence, in the light of both the palaeobotanical record we adduce here and independent data sets from other areas of expertise. I have deliberately kept quite separate in individual chapters the factual palaeobotanical record, the vegetation reconstructions, the set of palaeoecological hypotheses and my own conclusions, as well as some final general ecological considerations. It seems important not to trammel up the factual record and the ensuing interpretive steps with the biases and limitations of one observer.

I am no 'compleat palaeoecologist,' if such a scholar exists, so I will try to avoid straying far beyond the intellectual confines of my discipline (plant ecology). It is unfortunate but true that, no matter what Daedalian feats one may contrive, the separate branches of knowledge that comprise Quaternary science are advancing so rapidly as new methods and discoveries accumulate that an individual is unlikely to command enough comprehension of more than one major field at any one phase of his or her career to contribute usefully. There may be exceptions, but too many unhappy instances of ill-informed dabbling beyond a professional bailiwick clutter up the literature of our science.

Why the strictly Canadian region of eastern Beringia, it might be asked? Why not include Alaska and make a grand conspectus? The reason is straightforward. It so happens that the developments in Quaternary plant ecology in Alaska and in northwest Canada are out of phase. The beginnings were in Alaska, with Livingstone (1955a), Colinvaux (1964), Péwé (1965), Guthrie (1968), and others, culminating in Hopkins's (1972) imaginative synthesis. Efforts in Canadian Beringia were influenced by these notable advances and although some work began in the Mackenzie area in the sixties (Mackay and Terasmae 1963), the chief momentum

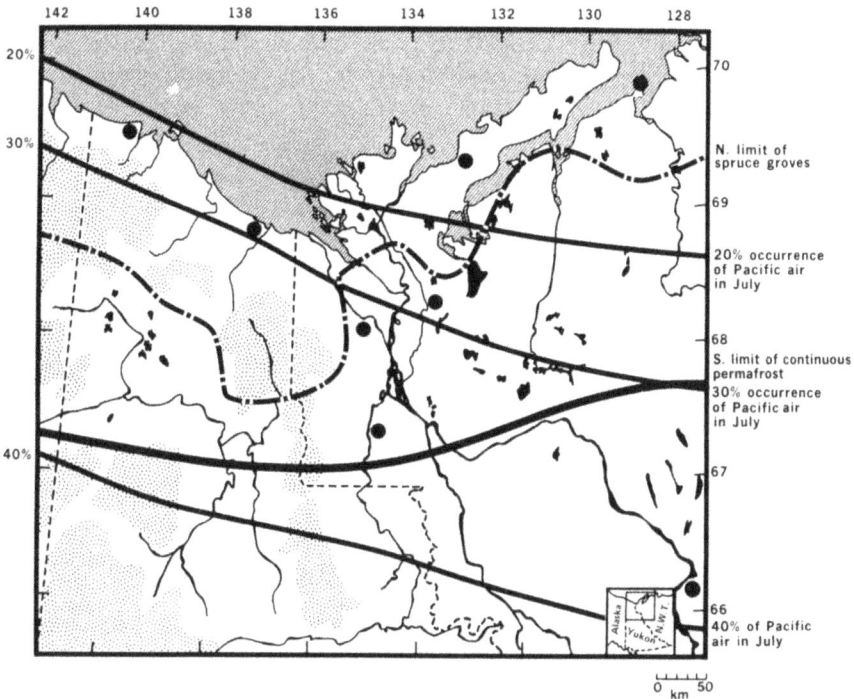

Figure 1 A sketch map of the study area showing the locations of the weather stations (solid dots). From north to south: Nicholson Point, Komakuk Beach, Tuktoyaktuk, Shingle Point, Inuvik, Aklavik, Fort Macpherson, and Fort Good Hope. The dashed and dotted line separates the arctic tundra region from the subarctic-boreal woodland region, and the limits of permafrost and air mass regions are shown.

developed in the seventies. Now we are ready, at least in plant palaeoecology, for a first summary, while our colleagues in Alaska have moved on to the second phase of research, using the new approaches and techniques that we have found so effective in addressing the problems raised by Hopkins (1972). It is highly likely that the 'new insights' intimated by Brubaker et al. (1983) will have yielded the basis for a compendium on Alaskan historical plant ecology in four or five years.

In the first few chapters I will describe in brief the physical setting – climate, geology, periglacial phenomena, and microtopography – and give a concise review of modern floristics and vegetation before we examine the palaeobotanical record, in outline from the Tertiary, in detail from the late Quaternary. The concluding chapters will address problems of vegetation reconstruction, palaeoenvironments, and various ecological implications.

The geographical limits of the area of interest can be seen in Figure 1. They consist of the Beaufort Sea coast from the Alaska border eastwards to Baillie Islands at longitude 128°w, the Yukon-Alaska boundary southwards to an arbitrary point between 65° and 66° N; the southern limit runs east-west between the Mackenzie Mountains foothills and the Peel River, joining in a right angle with the north-south limit a short distance south of Fort Good Hope. The approximate total area is 220,000 km^2.

2
The Physical Setting

1. Climate

Polar climates contrast sharply with those of temperate latitudes both because of the low angle of the sun and the resulting low values of radiation, and because of the compensating effects of long summer days. As a result arctic climates have very distinctive seasonal distributions of day length and radiation. Arctic land surfaces as a whole have net radiation values less than 15 kly yr^{-1}, and the northern boreal forest and transitional forest-tundra zone have from 14 to 25 kly yr^{-1}. Values for southern Canada are approximately 40 kly yr^{-1}, while in our area of immediate interest they range from 20 kly yr^{-1} in the southwest to less than 10 kly yr^{-1} on the Tuktoyaktuk Peninsula (Table 1).

The climate is dominated by the interplay of two contrasting atmospheric systems: the moist, Pacific maritime air masses controlled by the main planetary westerlies, and the arctic atmosphere to the north, which is often described as a vast cap of cold, low-energy air. The interactions between these systems are governed both by the annual (seasonal) north-south migrations of the course of the westerlies, and by the severely dissected mountainous terrain of Alaska and the Yukon (Hare and Hay 1974).

1.1. CYCLONIC AND ANTICYCLONIC PATTERNS

The intensity and frequency of synoptic-scale disturbances determine the ecologically important climatic characteristics of a region. In the arctic-subarctic these disturbances are caused primarily by the interactions between the arctic air masses of the polar region and the planetary westerlies of mid-latitudes. The pattern changes seasonally.

In winter, low-pressure westerly disturbances (cyclones) lie to the south of our area, travelling along the periphery of the large cold arctic vortex. Occasionally, blocking high-pressure systems deflect the course of these westerlies for short periods, causing anomalous periods of high temperature, reaching melting point for very short periods. The northern coasts rarely experience these breaks in the winter cold, the air temperature remaining at about –30°C with moderate east winds. These conditions change rapidly in spring when the arctic cap retreats northwards. The long days of April and May, with monthly bright sunshine hours in excess of 300, among the highest values found in Canada, cause a rapid rise in air temperature, particularly in the interior part of our region.

Summer circulation is dominated by the passage of many cyclones. In other words, our area is traversed by a frontal zone marking very roughly the boundary between the northern arctic mass and the warm, moist Pacific system. The arctic air mass expands to the south again in early October, and by the end of that month our area falls entirely within its influence.

1.2. REGIONAL CLIMATES

As a consequence of the seasonal interactions between the two major circulation systems we can identify two regional climates. It must be stressed that they intergrade, however, and, especially in light of the sparse available climatic data base, they should be regarded as broad categories recognized for descriptive convenience. These are the *continental subarctic climate*, which characterizes all of our area except the north coast and the Tuktoyaktuk Peninsula, and which is distinguished by early springs, and summers largely influenced by the cyclonic activity of the frontal belt; and the *arctic coastal climate* with late springs and a prevalence of arctic air in summer.

Let us examine each of these regional climates in more detail.

The task of assembling an account of the climate of the far northwest has been greatly eased by the recent compilations of Burns (1973, 1974). He has brought together in an attractive format, with many tables, diagrams, and maps, a detailed review of the subject. As he points out, two of the inescapable limitations are that large segments of the area lack weather station and that the existing network is so sparse and of such a short period of operation (maximum from 1941 to the present) that detailed adjustments of temperature records to account for orographic variations are not feasible. As a result, the summaries and generalizations made here should be used only at a broad, macroclimatic scale. However, many ecological adaptations to arctic and subarctic environments involve the exploitation and use by animals and plants, and probably by people, of local habitats with particular meso- or micro-climates. It is increasingly recognized, as studies of arctic microclimates proceed, that ecologically highly significant differences between sites varying in local relief and topography result from the low maximum angle of the sun. Also, an important characteristic of arctic winters in mountainous terrain is the entrapment and drainage of cold air. We will refer to examples of these local environments in later chapters, and it will be obvious already that this limitation on our knowledge of modern bioclimates will severely restrict our capacity to make detailed palaeoclimatic reconstructions.

In addition to the compilations by Burns (1973, 1974), the detailed summaries published by the Atmospheric Environment Service of Canada for the eight weather stations in our area have been consulted. The positions of these stations are shown on Figure 1; they are at Fort Good Hope, Fort McPherson, and Aklavik, all of which have records for the 25–30-year period preceding 1970, and at Inuvik, Komakuk Beach, Shingle Point, Tuktoyaktuk, and Nicholson Point, where only 10–14 years of record exist up to 1970. It will be clear from the distribution of these stations that data are lacking for the large southwest segment, so I have included the 30-year records from Dawson to provide representation. Dawson lies approximately 140 km to the south of the southwest corner of the area, at 64°04′N, 139°26′W, 300 m asl.

(a) *Continental Subarctic Climates*
These characterize the area lying south of a line passing along the northern foothills of the British Mountains and trending east-northeast along the axis of the Eskimo Lakes (Figure 1). The main ecologically important aspects of the climate are summarized in Table 1, and Figure 2 provides a graphic summary of the relatively steep gradient that separates this zone from the arctic coastal zone when important thermal and precipitation factors are examined.

This subarctic zone has relatively high values for the important thermal parameters, net radiation 14–20 kly yr^{-1}, mean monthly temperatures for the period May to August ranging from 6.7°C to 12.7°C, an annual frost-free period of 50–90 days, and degree-days above 5°C in the range 650 to about 1,000. Also, both snow depth values (1–2 m) and total estimated effective precipitation values are high. Winters are characterized by long periods of relatively calm, intensely cold air with a strong temperature inversion. Although we lack specific records from the interior of the North Yukon, it is clear by extrapolation from adjacent Alaska, as summarized by Barry and Hare (1974), that locally extreme cold occurs in mountain areas because of entrapment of calm air and cold air drainage into wide valleys with gentle gradients. Rare occurrences of incursions of much warmer Pacific air, caused by blocking anticyclones, produce short periods (a few days) of near thawing conditions. Snow cover is deep, and shows lower densities than in the arctic areas and minimal crust development.

The salient, distinguishing feature of this zone is the relatively early spring, with rapid warming, and thawing of snow and lake ice. During the period from April to June, along a roughly west-southwest to east-northeast gradient, day lengths and monthly sunshine totals of 18 and 300 hours respectively transform the landscape from the cold snow-clad aspect of winter to one of spring, with rapid melting and rapidly increasing air temperatures. This contrasts sharply with areas east of the Mackenzie Valley, where spring arrives one month later.

In summer, high frequencies of occurrence (30–50%) of Pacific air in the form of travelling cyclones, the westerlies, produce increases in cloud and precipitation (rain) and mean July temperatures of approximately 15°C.

Table 1 A summary of the climate, based on data in Burns (1973, 1974)

	Nicholson Pen.	Tukto-yaktuk	Komakuk Beach	Shingle Point	Inuvik	Aklavik	Ft McPherson	Ft Good Hope	Dawson
Mean annual global solar radiation (kly yr⁻¹)	76	77	77	78	78	78	81	80	85
Mean annual net radiation (kly yr⁻¹)	8	10	13	14	14	15	16	16	20
May-June-July-August mean daily air temperature (°C)	3.3	4.8	2.8	4.9	6.7	8.4	10.0	11.5	12.7
January mean daily air temperature (°C)	-29	-27	-24	-25	-28	-29	-29	-32	-29
Mean annual daily temperature (°C)	-12	-11	-11	-10	-10	-9	-9	-8	-5
Mean date of first frost	30/VII	10/VIII	30/VII	10/VIII	15/VIII	30/VIII	30/VIII	20/VIII	30/VIII
Mean date of last frost	5/VII	2/VII	8/VII	20/VI	25/VI	18/VI	5/VI	30/V	30/V
Mean annual frost-free period (days)	25	50	28	45	50	70	82	85	90
Degree-days above 5°C	233	372	165	386	654	656	766	927	1,015
Frequency of Pacific air in July (%)	6	15	20	25	28	30	35	38	50
Mean date lakes clear of ice	5/VII	5/VII	2/VII	5/VII	20/VI	25/VI	25/VI	20/VI	20/VI
Mean date lakes completely frozen	5/X	5/X	5/X	5/X	10/X	10/X	10/X	15/X	20/X
Mean maximum lake ice thickness (cm)	185	180	180	175	170	170	170	165	150
Estimated total precipitation, evaporation + run-off (mm)	203	229	230	304	317	380	405	450	620
Snow depth (mm)	380	560	520	830	1,750	1,400	2,200	1,375	1,360

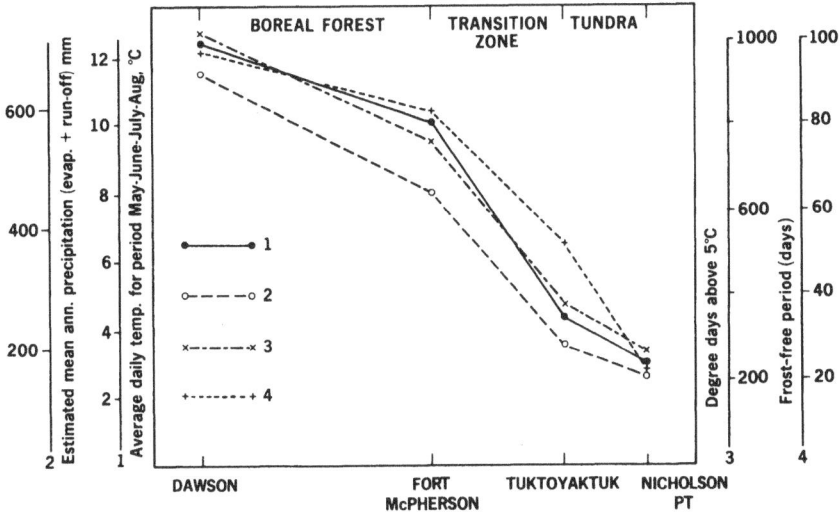

Figure 2 Profiles of various thermal and precipitation parameters along a transect from Dawson to Nicholson Point that traverses the study area in a roughly south-southwest to north-northeast direction.

(b) *Arctic Coastal Climates*

These occupy the northeastern segment (Figure 1) and contrast sharply with the subarctic regime, particularly in spring and summer. At the boundary with the continental subarctic zone the values of most thermal factors are sharply depressed. The transition coincides roughly with the ecological transition from forest to tundra. We have already noted the contrast in mean annual net radiation; this zone, in common with the entire arctic, has values less than 15 kly yr^{-1}. The mean monthly air temperature values of this zone do not differ much from those of the subarctic continental zone during fall and mid-winter months, but they are markedly less for late winter, spring, and mid-summer. A comparison of these values from Fort McPherson and Tuktoyaktuk shows the differences (Figure 3). In general, the summer months, May to August, have a mean daily temperature for the entire period of 3–5°C. Degree-day values above 5°C for the entire year range from only 200 to 400, and we should note that 95% of these occur between 1 May and 31 August, almost one-third of the values recorded in the subarctic zone. Similarly, the frost-free period is very short, from 25 to 50 days in length. Total precipitation and snow depth values are less than half of those for the subarctic zone, and the density and frequency of ice crusts are significantly greater (Mackay et al. 1975).

Winters are long, with intense cold broken only rarely during the period from October to April. The daily mean temperature remains below 0°C for approximately 40 weeks. As a result of the relatively flat, treeless, snow-covered winter

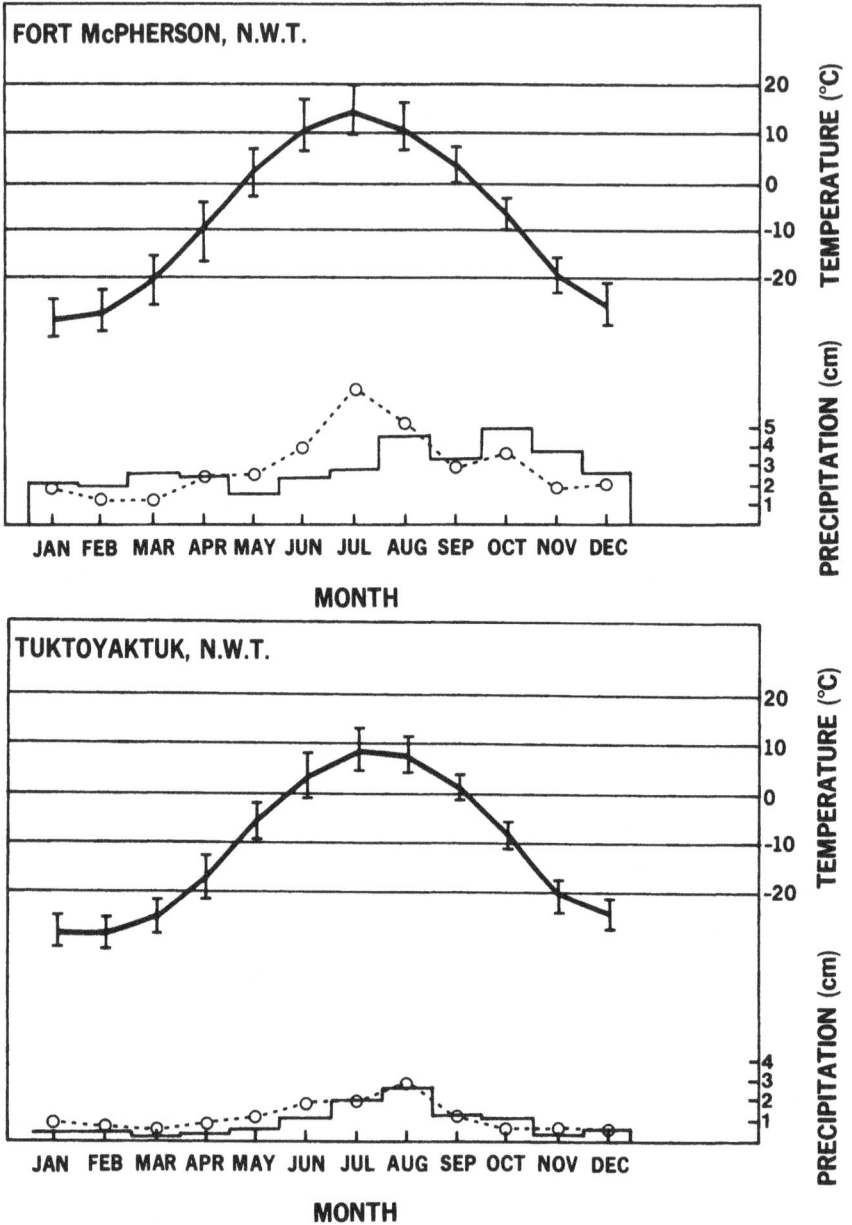

Figure 3 The temperature and precipitation means and the temperature extremes for a subarctic (Fort McPherson) and an arctic (Tuktoyaktuk) station, adapted from Burns (1973, Figures 4.38 and 4.40). The histograms show the mean monthly total precipitation and the open circles the monthly maximum precipitation in 24 hours.

landscape, cyclonic breaks in the winter dominance of arctic air are rare, and the high spring albedo of these surfaces effectively delays the incursion northward of the Pacific air masses. The result is that large lakes on the Tuktoyaktuk Peninsula still have ice at the end of the first week of July, and the last frost occurrence is rarely before that time. The summers are short, cool (mean July temperatures from 7° to 10°C) and cloudy, partly because of the proximity of a cold sea.

Many of the important ecological characteristics of northern climates are not registered in regional, macroscale meteorological information, as Mackay and Mackay (1974) emphasize: 'Differences in absorption of solar radiation, energy emission, slope and aspects, heat capacities and conduction, and in local energy transfer mechanisms create meso- and micro-climates which are biologically very important.' We shall return to these questions towards the end of the book when palaeoenvironmental reconstructions are considered, but one final topic, also not obvious in tables of climate summaries, must be stressed. It is well known (e.g. Hare and Ritchie 1972) that snowfall and snow-cover regimes are of great ecological importance at high latitudes but equally the difficulties of accurate measurement are formidable mainly because 'snow does not lodge readily in windy sites, such as climatological stations on open land. Moreover, the available snow gauges are of dubious performance' (Hare and Ritchie, p 362). A recent study in the central Canadian arctic (Ohmura 1982) has illuminated the important role of snow cover, particularly as the operative factor in distinguishing the spring season in the two climatic regions that occupy our area. In spite of lower total amounts of snow in the arctic, the result of redistribution and compaction by wind is a dense snowpack which melts slowly. In May and early June, before snowmelt begins, global radiation is at its maximum but albedo values of 0.8–0.6 keep net radiation totals negative or near zero. When melt occurs, albedo drops sharply to about 0.2 in the tundra and the energy balance of the surface changes drastically, giving positive net radiation totals. The short, relatively mild arctic summer begins – in July – and mean net radiation values of about 10 MJ m^{-2} d^{-1} are reached, comparable to those of southern Canada. By contrast snow in the open boreal woodlands of the subarctic region is rarely redeposited by wind, and stays light and fluffy. In early spring (May), with maximum radiation values, the low albedo of the northern boreal zone, largely because of the large absorption characteristics of the dark green spruce trees, results in a rapid increase in surface net radiation (+14.5 MJ m^{-2} d^{-1} in May, with albedo 0.2; after Ohmura 1982, Table v). Snowmelt occurs immediately.

Many authors have referred to the importance of snow cover to animal ecology. For example, Hoffmann (1974) elaborates on the themes that 'small mammals form patterns of dispersion in response to the patterning of winter snowdepth,' and 'larger herbivores are dependent on the spatial distribution (of snow) and also its quality "for mobility and foraging capacity."' We will discover that this is a significant but still poorly understood palaeoecological factor.

2. Geology

The physiographic characteristics of the area are the result of a complex array of features of bedrock geology, variably modified by Quaternary deposits and processes. A detailed examination of the structural geology would not be appropriate in this book, but it is important to understand in outline the basic framework of tectonics and structures. Interested readers can find detailed accounts and additional references to literature in articles by Yorath et al. (1969), Norris (1973), and Norris and Yorath (1981). The following short summary draws heavily on these sources.

2.1. STRUCTURAL GEOLOGY

Our area is the meeting-place of three important North American geological provinces: the Arctic Continental Shelf, which extends along a broad coastal belt of the northern mainland of western Canada and Alaska; the Interior Platform, an extensive system of sedimentary rocks extending in a tongue north from the southern borders of the prairie provinces, between the Shield and the Cordillera, to the Mackenzie Delta; and the Cordilleran orogenic system comprising the vast western montane north-south axis of the continent (Figure 4).

(a) *Arctic Continental Shelf*
This feature, of growing economic significance, consists of a shallow homocline (a unit or series of rocks with the same dip) overlain by thick deposits of unconsolidated Quaternary and recent material. The rocks of Tertiary and greater age do not form outcrops, and the modern coastal plain features are largely uninfluenced by the structural and tectonic geology. Offshore, the shelf area is a gradually declining plain, 50–200 m deep, with thick uniform marine deposits interrupted only by the Mackenzie Canyon north of Herschel Island, and various submerged pingo-like features (Shearer et al. 1971). The northeast-southwest alignment of the Eskimo Lakes and Liverpool Bay is interpreted as an extension of a tectonic feature, the Aklavik Arch (described below). However, the bedrock is buried by thick deposits of Pleistocene fluvial and deltaic silts, sands, and gravels.

(b) *Interior Platform*
The northern extremity of this large province forms the structural basis of the southeast segment of our area (Figure 4). It is represented by the Anderson Plain, a gently west-dipping homocline, made up of Palaeozoic, Cretaceous, and Tertiary rocks that outcrop locally and control the drainage patterns in extensive areas where the glacial drift mantle is thin and patchy.

(c) *Cordilleran Orogenic System*
This complex system consists of four basic elements: platforms, fault arrays, depressions, and uplifts. The grouping and mapping presented here is taken

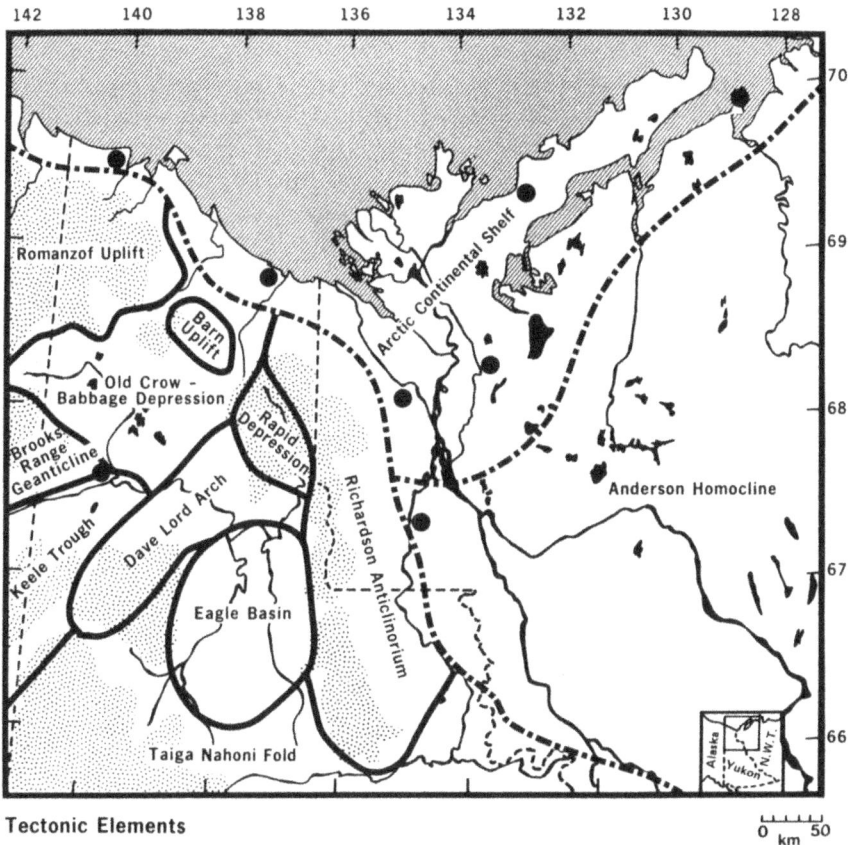

Tectonic Elements

0 km 50

Figure 4 A simplified map of the main tectonic regions. The dashed-dotted lines separate the three main geological provinces: the Arctic Continental Shelf; the Interior Platform, represented here by the Anderson Homocline; and the Cordilleran complex, which is subdivided into 10 tectonic elements.

directly from Norris and Yorath (1981). Platforms consist of undeformed, layered sequences of sedimentary rocks, usually gently inclined and stable, forming parts of the craton (the large, relatively immobile parts of continents). Fault arrays are bundles of roughly parallel faults, and they are common in our area in, for example, the South Richardson Mountains. Depressions and uplifts are related, complementary features: the former are structurally negative, the latter positive. They are the result of episodes of uplift and depression, variably altered by deformation, erosion, and sedimentation. There are 10 principal tectonic elements in our area (Figure 4).

The *Romanzof Uplift* consists of Palaeozoic and Proterozoic rocks, uplifted, folded, and buckled mainly in late Cretaceous and Tertiary times. Adjacent to it, to the east and south, is the *Old-Crow-Babbage Depression*, a complex of Proterozoic to Lower Cretaceous sedimentary rocks with local, small granitic intrusions.

Large parts are buried beneath thick Quaternary deposits that form the modern Old Crow Flats. To the west of the Old Crow Flats is the eastern edge of a montane feature known as the Brooks Range Geanticline, represented in our area by the *Old Crow Mountain* complex, consisting of intrusive, mainly granitic rocks of Proterozoic age. To the east and northeast of the Babbage Depression is the *Barn Uplift*, an acutely folded and faulted complex of shales, cherts, quartzites, siltstones, and limestones of Lower Palaeozoic age. To the south is a structural low, the *Rapid Depression*, a northeast-trending feature of siltstones, sandstones, and conglomerates of Cretaceous age.

The *Aklavik Arch* is a northeast trending structure that extends from Bell Basin across the Mackenzie Delta to near Sitidgi Lake. It includes the *Dave Lord* and *Campbell Uplifts*, and extends from the *Keele Range* in the southwest of our area. It consists of Proterozoic to Tertiary sedimentary rocks. The *Eagle Basin* is a deformed foldbelt of Palaeozoic and Mesozoic siltstones, slates, and other clastic strata. The *Richardson Anticlinorium* (a series of anticlines and synclines that form a general arch or anticline complex) is a fault array of nearly vertical structures, running north-south for about 1,000 km from the Mackenzie Foldbelt south of the modern Peel River to the Beaufort Shelf. It consists of Proterozoic to Tertiary rocks that were tectonically active between the Late Cretaceous and the Early Tertiary.

This complex structural base has been modified by a great variety of Pleistocene and recent processes. The result is a set of physiographic units, defined in terms of geomorphology, that provide a useful framework for describing and mapping the physical features of the area.

First, however, we must trace the Pleistocene geological sequence of events that exercised such a major influence on the geological features of the area.

2.2. PLEISTOCENE GEOLOGY

There remains much uncertainty about the extent and timing of Pleistocene glaciations in this area, and the following account should be regarded as tentative.

The salient feature is the clear separation of the area into an eastern glaciated and a western unglaciated part (Figure 5). The differences in the landscapes of the two areas are striking. To the east of the Richardson Mountains axis, there are extensive areas of hummocky moraine, glaciofluvial deposits, glacially fluted hills and ridges, eskers, and other features of glacial origin. To the west, there are extensive pediment surfaces of colluvial silts and gravels, exposed bedrock, and a complete absence of landforms associated with past glacial cover.

Although the details remain unclear, there is no disagreement among geologists (see, for example, Hughes 1972 and Rampton 1982) that continental Laurentide ice extended from its eastern centre of accumulation no further than a line running, roughly, along the eastern slopes of the Richardson Mountains and then northwesterly parallel to the Yukon coast to a point near the coast southwest of

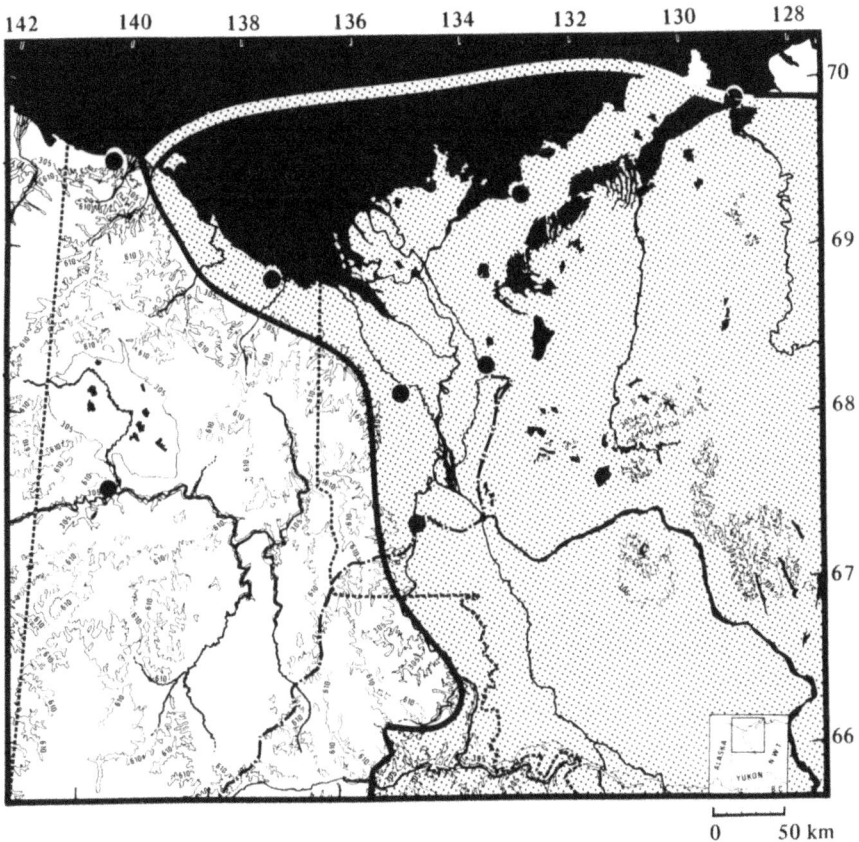

Figure 5 At the maximum extent of Wisconsin glaciation the Laurentide continental ice extended westward to the limit shown approximately on this sketch map. The time of this event is uncertain, as discussed in the text.

Herschel Island (Figure 5). The area to the west of this line was unglaciated, with the exception of valley glaciers that occurred in the Ogilvie, Mackenzie, and Selwyn Mountains, at and beyond the southern limit of our area of study. There is in addition evidence of minor valley glaciers in the Richardson Mountains that were probably confluent with Laurentide glaciers.

The primary evidence for glaciation is of two types: first morainal features and drift erratics that provide direct records of the extent of cover and limits of glaciations and, second, stratigraphic evidence found in exposed sections or drilling records. Unfortunately, persistent gaps and uncertainties in both types of data for this region hamper the task of developing a secure account of the glacial and related events.

The early account by Hughes (1972), which was offered as a quite tentative, qualified working hypothesis, rapidly became the established dogma as research in Quaternary studies intensified in this part of Canada.

This version suggests that the maximum extent of Laurentide ice marked the oldest glacial event in the area, probably of Wisconsin or Pre-Wisconsin age. This ice sheet caused a diversion northward of the upper Peel River drainage into the interior plains region. The diverted watercourse entered what is now the headwaters of the Eagle River by a large canyon, currently occupied by Davis Lake (Plate 1). Similarly, the ancestral Porcupine River, which had flowed to the northeast through a gap in the Richardson Mountains into the Mackenzie drainage, was diverted to the west by ice blocking this valley. The result of these diversions was the formation of three basins of accumulation – glacial lakes – in the three areas of lowland now referred to as the Bell Basin, the Old Crow Flats, and the Bluefish Basin. The stratigraphic and chronological relations of these basins are poorly understood, so it would be unwise to assume that their periods of high water were contemporaneous. However, it does appear that two major episodes of glacial lake sedimentation, presumably coeval with Laurentide glacial maxima, are recorded in exposed sections of these basins. They are referred to informally as the Upper Lake and Lower Lake units, and they are separated by complex, poorly understood sequences of fluvial, lacustrine, and deltaic deposits. Jopling et al. (1981) designate these units, as they occur along the Old Crow and Porcupine rivers (in the Old Crow and Bluefish basins respectively), as the Glacial Lake Kutchin and Glacial Lake Old Crow units. There is general agreement that the upper unit is of late Wisconsin age (i.e. roughly 30,000–18,000 yr BP) while the lower is referred to as 'Illinoian,' somewhat in excess of 120,000 yr BP (Jopling et al. 1981; Morlan 1980).

A second limit to Laurentide ice expansion is shown in Figure 6, following Hughes (1972), but it has been difficult to trace completely. One working hypothesis by Rampton (1982), which we will follow here, is that the ice limit mapped by Hughes (1972) along the lower east flanks of the Richardsons is continuous with the moraine sequence identified by Hughes (personal communication) as the Tutsieta moraine (Figure 6), mapped in the Travaillant Lake and Tutsieta Lake regions of the southeast sector of our area. However, it is not clear how this correlates, if at all, with any stratigraphic units recorded in the North Yukon. Interested readers will find Hopkins's (1982) attempt to develop a consistent stratigraphy for the entire Beringian region of value, but these questions of the age and sequence of events of Quaternary history will be examined in more detail in chapter 5. From our present perspective – physiography – problems of chronology are relatively insignificant.

3. Permafrost and Periglacial Features

The area displays a remarkable range of surface features that are usually described as patterned ground or periglacial phenomena. They are expressions of

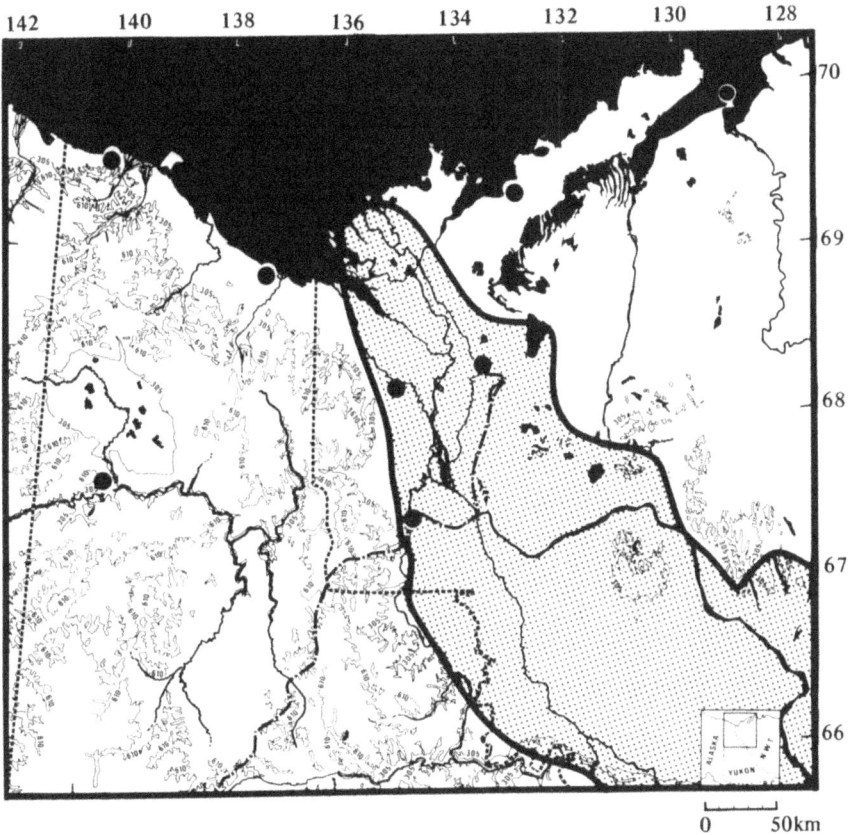

Figure 6 The shaded area was occupied by a tongue of Laurentide glacial ice about 14,000 yr ago, but the reader should consult the text to appreciate the uncertainties in our understanding of the chronology of Pleistocene glaciation in the region. The eastern limit of this lobe is defined by the Tutsieta moraine, after Hughes (1980, personal communication); otherwise the ice limit position is based on Map 31, Geological Survey of Canada, 1979, and Prest, 1970.

the effects of the climate on the structural and physiographic materials, with the result that there is both an approximate north to south trend in the types of patterned ground and a clear relation to the regional physiography. They are microtopographic manifestations of processes that are directly or indirectly due to the following factors, singly or in combination: the presence of permafrost, frost action, ice segregation, soil displacements, differential slope movement, local winds, and influences of local slope aspect. The scale of these phenomena ranges from the average size of mature pingos (~150 m basal diameter, 40 m height) to that of earth hummocks (1 m basal diameter, 40 cm height). There is a large literature on these topics, both in general (e.g. French 1976; Washburn 1973) and with reference to this area (Mackay 1972, 1979; Zoltai and Tarnocai 1974, 1975; and many others, throughout this chapter), and this reflects both their intrinsic interest and complexity and their critical relevance to many aspects of northern industrial development. The concise review that follows is included because of their importance in plant ecology and their significance in Quaternary palaeoecology.

3.1. PERMAFROST

Permafrost is defined as subsurface material that remains frozen throughout a period of several years – in other words, perennially frozen ground. Where such terrain prevails everywhere, except below and near oceans and large lakes, the area is described as being in the zone of *continuous permafrost*. In practice, this is the area north of a line delimiting the –5°C ground temperature isotherm (Brown 1967), which passes through our region (Figure 1) roughly along latitude 67°50′N. South of this line lies a broad belt where permafrost is patchy – the zone of *discontinuous permafrost*. We should note that the –5°C ground temperature datum, in common with all ground temperature standards, is measured at that depth where annual variations do not occur – in this region, at about 15 m below the surface.

Breaks in permafrost zones, known as taliks, occur beneath large lakes and ocean margins. Otherwise, permafrost thickness varies from 50 m near the boundary of the continuous and discontinuous zones to greater than 1,000 m in parts of the arctic. Depths of 600 m have been recorded in the vicinity of the Mackenzie Delta (Mackay 1967), and we should note that the depth of permafrost is often directly related to the regional history of glaciation. It is well known that glaciation in Siberia was relatively localized and sparse whereas in North America it was extensive and thick. This accounts for the extensive development of periglacial phenomena and deep permafrost in Siberia and their relative scarcity with shallow permafrost in much of glaciated North America. Similarly, a long history of ice-free arctic-subarctic conditions explains the thick (>600 m) permafrost in unconsolidated sediments on the Yukon and Northwest Territories coastal plain (Mackay 1971; Mackay et al. 1972). Roughly 100,000 years are required for the growth of such a depth of permafrost in these materials, assuming conservative lowering of ground temperatures during the latest glacial cycle (Sherbatyan 1974).

The zone of substratum, immediately below the surface, that freezes and thaws each year is referred to as the active layer. Its thickness varies according to the thermal properties of the surface materials, the vegetation cover, the slope and aspect, and the soil moisture conditions.

The seasonal freeze-thaw cycle of the substratum in permafrost areas involves a complex interaction of physical processes that produce varied, often poorly understood patterns. Permafrost can act as an effective block of downward drainage of soil water released in the active layer during summer. As a result, slope movements of various types – mass wasting, soil creep, solifluction – are common.

3.2. PINGOS

Pingos are topographic features restricted to the permafrost zone. In our area they are concentrated in the coastal lowlands of the Tuktoyaktuk Peninsula, adjacent Richards Island, the distal part of the Mackenzie Delta, and the coastal lowlands of the Yukon (Porsild 1938). They are ice-cored, roughly conical hills (Plate 4) varying in size from 60 m height and 300 m basal diameter to small mounds of less than one-tenth those dimensions. They are widely distributed in circumpolar arctic-subarctic regions, and there are several types with different modes of origin. As it happens, all the pingos in our region are of the 'Mackenzie' or closed-system type and they have been studied in detail by Mackay (1979 and many other papers cited in that article); their mode of formation is reasonably well understood. They are formed by permafrost aggradation in sandy, supersaturated sediments, resulting in pore water expulsion and the establishment of a localized increase in hydrostatic pressure. This occurs in partially or completely drained lake basins, where a drop of water level causes the lake to freeze to the bottom in winter. In this area, the drainage event that triggered pingo growth appears to have been the local, catastrophic collapse of frozen soils, lowering lake levels rapidly. The subsequent permafrost aggradation often results in the development of a localized hydrostatic pressure causing a doming upwards, thrusting up the lake sediments overlying a core of ice. The resulting mound, a pingo, grows upwards and outwards, initially rapidly (~1.5 m/yr), later with sharply decreasing speed, reaching approximate stability after several centuries.

All stages in the growth, maturation, and collapse of pingos can be seen in the roughly 1,500 examples that occur in the coastal lowlands of our area (see Plate 4). Collapsed and eroded pingos are of palaeoecological interest; in relict form they can be useful indicators of past periglacial environments (e.g. in northwestern Europe, Pissart 1963; Mitchell 1971; Watson 1971). We have analysed the exposed pond sediments of collapsed and eroded pingos to reconstruct past vegetation and environment (Hyvärinen and Ritchie 1975), as noted in chapter 5.

3.3. ICE-WEDGES

A second common surface feature is patches of polygonal terrain (Plate 6). They are variously described as ice-wedge polygons, polygon nets, tundra polygons, and high- and low-centred polygons. In our region they are abundant wherever poorly drained, fine-grained materials occur; their areas of maximum concentration are the Coastal Lowlands, the Peel Plain Lowlands, and the Old Crow, Bluefish, and Bell basins, but they occur in every physiographic unit. They are more widespread than pingos, occurring in both the continuous and discontinuous permafrost zones. Mackay (1972) estimates that their total area in the Northern Hemisphere is approximately 2,600,000 km^2.

Their geometry and dimensions vary. The most common reticulum type in our area is roughly hexagonal, with the ice-wedge ditches 1–2 m wide and the intervening areas 30–60 m wide. Locally, usually adjacent to large extant or former lakes (e.g. in the Old Crow Flats), the wedges form an orthogonal or rectangular pattern, with one set of wedge ditches normal to the lakeshore or former beach and the others set at 90°.

The mode of formation of these features has been studied in detail, for example in our area by Mackay (1972). They require permafrost terrain and a mean annual temperature of –6°C or cooler. They occur frequently in the zone of discontinuous permafrost (Zoltai and Tarnocai 1974, 1975), but they are generally not growing actively. Their formation begins as a process of contractions in ice-rich, frozen terrain. This surface cracking, in response to low winter temperatures, forms the initial reticulum or network of fissures. The cracks grow by downward percolation and in-filling by spring melt-water, to form an ice-wedge or seam. The fissure reopens in subsequent years and the process continues, producing a wedge-shaped, icy unit (Plate 7). The pattern of seasonal cracking and wedge growth is varied, and wedge width growth rates of the order of 1–2 mm per annum appear to be normal (Mackay 1973). Ridges develop along the wedges, often supporting a willow–dwarf birch vegetation in contrast to the sedge–cotton-grass cover of the intervening areas, and this results in a highly visible pattern when viewed from the air or on air photographs.

Ice-wedge fields of two types occur widely: high-centred and low-centred. In the former, the ground between the wedge ditches becomes domed often with relatively dry, humous peat near the surface supporting a characteristic heath-lichen cover. Low-centred polygons have wet centres, slightly lower than the wedge ridges, often with pools of open water or sedge–cotton-grass vegetation over saturated sedge-moss peat.

In southern parts of our area, for example in the Bluefish Basin immediately south of the village of Old Crow, late stages of apparent polygon decay can be seen. These are the result of thermokarst – melting of ground ice. It appears to occur at the junction of wedge ditches, forming elongate ponds; and in areas with slight slope, the original polygon pattern appears to have degenerated into a series of roughly parallel elongate ponds and ridges. Similarly, thermokarst melting of low centres can lead to extensive ponding and disintegration of the original patterns.

Ice-wedge polygons are of interest to plant ecologists because of the pattern of communities associated with them, and to palaeoecologists because they are often preserved as informative palaeoenvironmental features in both exposed sections of Pleistocene sediments and as persisting surface patterns in areas now far removed from any active permafrost (e.g. Southern Ontario, Morgan 1972).

3.4. EARTH HUMMOCKS

These features, variously described as hummocks, frost boils, and non-sorted circles (Washburn 1973), are common throughout the area on level or gently sloping surfaces (1–2°), usually with fine-grained (silt, clay) parent materials. Farther south in the Mackenzie River valley they become the dominant patterned ground type (Zoltai and Tarnocai 1974). They consist of large patches or 'fields' of roughly circular domes, 1–1.5 m in basal diameter and 40–50 cm in height (Plate 8). They occur in both the tundra and forested regions. In the tundra they occupy large areas of spotted landscape, easily recognized from the air. In forested areas they occur as partially or completely open (treeless) patches, often on slightly sloping or level plateaus, each patch consisting of a field of hummocks, with a characteristic vegetation cover (Zoltai and Pettapiece 1973, 1974).

A related phenomenon, gelifluction (Washburn 1973), is the slow down-slope movement or flow of saturated sheets of surface materials as a result of impeded drainage in frozen terrain during the thaw season. Many types of such movement occur, depending on the surface materials, angle of slope, and other factors, and they can often be recognized as lobate sheets of soil material with a down-slope margin raised 50–100 cm above the surface. Sorting can occur in this general process, resulting in irregular striped terrain with stripes roughly parallel to the contours. These features are common in our area on sloping pediment surfaces in the Peel Plateau uplands of the North Yukon and the Northwest Territories.

Hummocks are the result of frost action in the substratum, often referred to as cryoturbation, but their precise mode of origin remains poorly understood. In section (Plate 8) they often show a domed permafrost table roughly parallel to the surface contours, and the centre, where upthrusting or injection has been active, usually has a concentration of finer-grained clay material. In exposed tundra areas the hummock tops are often denuded, and wind deflation in summer occurs. In wooded areas the hummock summit often supports a dense lichen-moss mat. In all areas, their presence is a reliable indicator that fine-grained mineral substrata are close to the surface, the depth of peat accumulation in the hummock interstices rarely exceeding 25 cm.

3.5. STONE CIRCLES, NETS, AND STRIPES

Upland surfaces that lack peat accumulation and where at least some glacial or residual soil has accumulated often have microrelief features in the form of circles, nets, or stripes of variously sorted or unsorted rocks, stones, and fines. These result from cryostatic pressures causing differential movement and sorting of the mineral substratum. In our area sorted and unsorted stone circles and stripes are more common in the western part, where upland pediment and montane surfaces prevail (Plate 5).

3.6. PALSAS AND PEAT PLATEAUS

In the southeastern section of the area, chiefly in the lowlands of the Mackenzie River valley south of approximately the 67th line of latitude, two types of patterned ground are common: palsas and peat plateaus. They are characteristic of lowlands in the discontinuous permafrost zone, and they have been described in detail from sites in the Mackenzie valley by Zoltai and Tarnocai (1974, 1975). They are of circumpolar distribution.

Palsas are low mounds, raised 1–6 m above the immediate surroundings, 10–14 m in diameter, consisting of a cap of peat overlying a domed, frozen mineral substratum. Layers of segregated ice often occur near the contact between the organic and mineral units. The surrounding mires are usually of the fen type, and the palsas occur as low islands of peat. Their mode of origin has not been explained adequately, but the general consensus is that it involves differential peat growth and localized ice segregation in the underlying mineral layers, in contrast with that of closed system pingos, which are formed by injection ice.

Peat plateaus are related features in that they occur in the same physiographic position – mire lowlands – and in the same geographical zone. Some authors (e.g. French 1976) suggest that peat plateaus might result from the coalescence of palsas. They are frozen peat deposits of variable size, from a few square metres in area to several tens of hectares, raised about 1 m above the surrounding lowland water table. The surrounding fens often lack permafrost, and this is also true of many palsas (Brown 1967). Zoltai and Tarnocai (1975) describe a fairly consistent stratigraphic sequence through plateaus, from a cap of *Sphagnum* peat, a layer of sedge-moss peat, sometimes an aquatic, lacustrine organic silt, and a basal mineral layer. Ice-wedge polygons are common in peat plateaus.

3.7. THERMOKARST PHENOMENA

The term thermokarst describes the melting of ground ice followed by collapse or local subsidence of terrain (French 1976, p 105). These phenomena are common in arctic landscapes where extensive tracts of saturated sediments occur, often of alluvial, lacustrine glaciofluvial origin, and they result in several distinctive topographic features, of which thermokarst lakes, slumps, and erosional scarps or banks are common.

The coastal lowlands and the interior basins of the North Yukon show an abundance of thermokarst lakes. These are shallow (1–2 m), small (2–5 ha) to medium (10–500 ha) in size, often occurring in extensive fields or groups with the ponds or lakes showing similar shape and lineation. The satisfactory explanation of these oriented lakes has been the subject of much investigation (e.g. Mackay 1963), but so far it remains elusive. These lakes begin as a local melting and collapse event in a lowland permafrost area, and expand quickly by melting and collapse of the shoreline at rates of 15–20 cm per year. Collapse and shoreline retreat is more active along certain lake margins than others (usually the down-wind shore), and the lake migration, shape (elliptical or rectangular), and the orientation of the long axes tend to be constant within particular lowlands. Mackay (1963) describes a large group of these lakes in the eastern part of the

Tuktoyaktuk Peninsula, all of which show an approximate north-south orientation of the long axis, but with a slight clockwise shift eastwards from a mean of N 05°E to 11°E. Although there remain uncertainties, the general conclusion is that these particular lakes are in equilibrium with conditions today, and that the shape and orientation are determined primarily by wind effects. There is less information available on the large concentration of oriented, migrating thermokarst lakes in the Old Crow Flats (Plate 3), but they are being studied currently.

The presence of extensive ice-rich, fine-grained sediments along the arctic coasts of the Yukon and Northwest Territories results in extensive coastal erosion, particularly in the tract of coastal lowland west of the Mackenzie Delta (Plate 27). Mackay documents the rate of retreat of the coastal scarp as 30–60 cm / yr.

An understanding of the thermokarst phenomena, particularly thaw lakes, is of great interest to palaeoecologists, chiefly because of the potential for errors of interpretation of such disturbed, mixed sediments.

3.8. SOILS

By contrast with temperate regions, the distinction between soil type and microtopography is not clear at high latitudes. The result is that agreed systems of describing, grouping, and naming arctic soils have been difficult to achieve. The prevalence of cryoturbation and other ground frost factors have made it difficult to apply the orthodox concepts of soil genetic classification. There is a large literature on this subject, dating from the 1920s to the present; Tedrow (1977, Chapter 7) has provided a useful review of these questions.

Orthodox soil classification is based on the central concept of a mature, stable soil profile, differentiated into recognizable horizons or layers in terms of colour, structure, texture, chemical and other characteristics, developed in time from an interaction of geological, climatic, and biological factors. In arctic regions instability of surface materials because of cryoturbation and other permafrost-related factors is widespread, and soils that have developed in undisturbed conditions over long periods are relatively rare.

In recognition of these difficulties, Zoltai and Tarnocai (1974) have proposed a new, tentative classification which recognizes at the outset that soils developed in a permafrost landscape are sufficiently distinctive to require a special system of description. The central concept of soil type has been broadened to include 'all elements of the microtopography in cycles less than 7 m in width.'

In practice, in our area there are four main soil types, if we use the term in this broad way. (1) Rocklands, characteristic of montane and other uplands where bedrock is close to the surface. Here there is a preponderance of coarse, rocky substrata with no development of a soil profile. (2) Mineral soils strongly affected by cryoturbation, often occurring as part of patterned ground complexes, common on all uplands except in the southwest, wooded sector. (3) Mineral soils developed in areas lacking significant cryoturbation. These are referred to in the National Soil Survey of Canada scheme as Brunisols (Atlas 1974) and are distinguished by moderate profile development into a narrow humus zone below the surface litter horizon, a B horizon characterized by colour and texture and suggesting moderate podsolization, and the underlying parent material. These soils are common on uplands under spruce in the southwest sector. (4) Cryic gleisols and fibrisols, characteristic of poorly drained mineral and peat substrata, respectively, are prevalent in permafrost areas, particularly throughout the coastal lowlands, the Mackenzie Delta, the Peel Plain Lowlands, and the interior basins of the North Yukon.

Interested readers can pursue the topic in more detail in Tedrow (1977), Zoltai and Tarnocai (1975), and other publications.

4. Physiographic Regions

The study area has been divided into 15 units in Figure 7; I have attempted to follow as closely as possible the subdivisions delineated and described by the three main authorities on the land-forms of the northwest: Bostock (1948), Mackay (1963, Fig. 37), and Rampton (1982). Also, in deciding the degree of subdivision, which is of course determined primarily by scale, emphasis is given to units of apparent ecological importance; in that respect the map and descriptions by Oswald and Senyk (1977) have proved to be helpful. Physiographic regions are defined here as landscape areas separable by the prevailing relief, patterns of surface materials, and drainage characteristics. The smallest unit mapped here is approximately 3,000 km^2 in area, so it is clear that many small, distinctive units are not recognized in this general account.

4.1. MODERN MACKENZIE RIVER DELTA

This magnificent feature, at the mouth of Canada's longest river, is one of the major fluvial ecosystems in the entire circumpolar region. It is unique to North America, matched in size and complexity only by the Lena River Delta in the central Siberian arctic. It occupies approximately 12,000 km^2 between the Richardson Mountains on the west and the Caribou Hills and Quaternary uplands to the east. The delta presents an intricate pattern of innumerable channels, lakes, point bars, flats, and levees (Plate 2), extending from about 67°40′N to 69°40′N, nourished by the Peel River on its west flank and the Mackenzie River from the south. The reader is referred to Mackay (1963) for detailed descriptions.

4.2. PLEISTOCENE COASTAL PLAINS

A large, narrow tract of land, interrupted by the Mackenzie Delta, extends across the entire northern edge of our region. It consists of flat or gently rising land, with

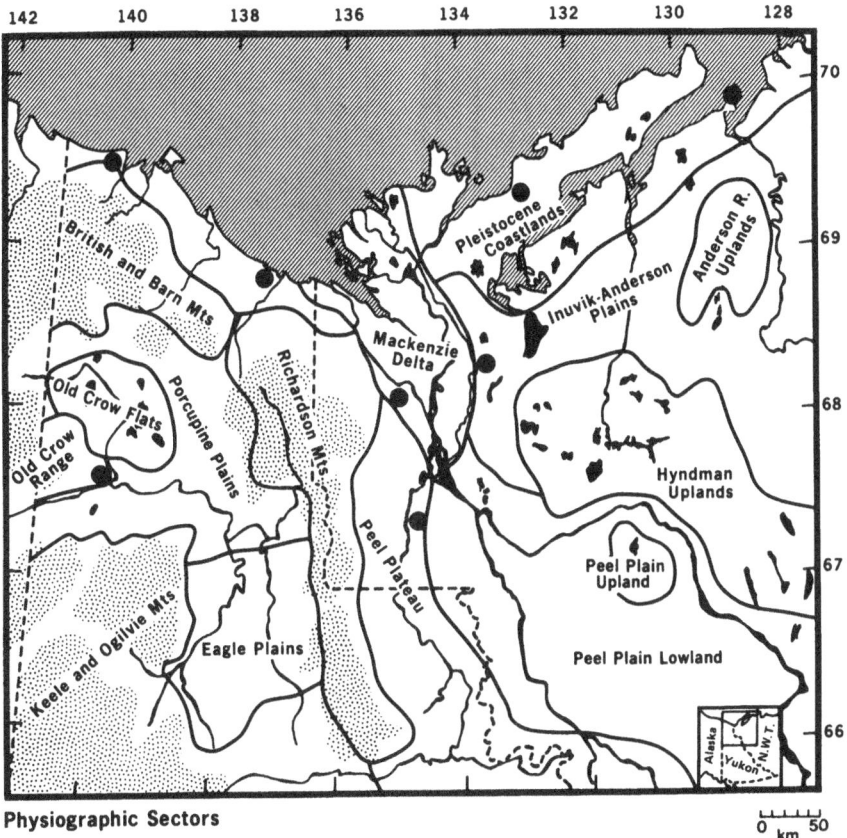

Figure 7 The approximate boundaries of the 15 physiographic subdivisions of the study area

relief not exceeding 60 m asl in the coastal part, rising to approximately 200 m inland toward the Yukon mountains. Along the Yukon coast there are large alluvial fan systems, at the mouths of the several rivers that empty into the Beaufort Sea. There is often a sharp erosional cliff along the coast (Plate 27) where landward erosion by melting of permafrost proceeds at the rate of a few metres per year (Mackay 1963). Elsewhere, associated with river-borne sediments, low coastal sand dunes are common. The interior is a flat, poorly drained surface with extensive ponding and marsh-mire development, giving way inland in the Yukon portion to a gently rising pediment surface. The Tuktoyaktuk portion of the coastal plains is distinguished by a greater density of shallow thermokarst lakes, of the oriented type on the distal portion of the peninsula, and a concentration of several hundred closed-system pingos in the central part of the peninsula (Mackay 1979). Ice-wedge polygon development is abundant. Parts of this unit, south of the settlement of Tuktoyaktuk near the Eskimo Lakes, are distinguished by hilly terrain, never higher than about 60 m, with irregular, deep (~10 m) lakes.

The origin of these hills, and of the finger-like ridges that separate the Eskimo Lakes from Liverpool Bay to the northeast, is unclear, but Mackay (1963) suggests they might be morainal. The mantle of unconsolidated sediment is thick, and it appears that some of it in the Tuktoyaktuk Peninsula area had its origin as fluvial-deltaic deposits from an ancestral Mackenzie River that flowed through the Campbell Lake graben near Inuvik and into the Eskimo Lake tectonically controlled channel by way of the wide channel occupied today by Sitidgi Lake. A ridge of low hills and irregular, moderately deep lakes in the central part of the Tuktoyaktuk Peninsula was originally described by Mackay (1963) as dead-ice moraine, but he later revised that to topography caused by extensive thermokarst activity (Mackay et al. 1972).

4.3. ANDERSON RIVER UPLANDS

This is a region of irregular, hilly terrain, with relief up to 400 m. It consists of thick Pleistocene deposits of unknown age, with many small lakes and deeply cut stream channels, sometimes occupying old melt-water channels.

4.4. INUVIK-ANDERSON RIVER PLAINS

This is an area of intermediate relief (40–300 m) with varied glacial features forming rolling, fluted, or irregular topography. The section with fluting occurs immediately to the west of Anderson River and extends east to the Kugalik River and southward in a wide swathe from the Crossley Lakes to Canot Lake. The terrain is marked by lower (~60 m), long (1–3 km) linear features trending roughly north-south. Elsewhere, fine-grained glaciofluvial materials mantle the landscape, with the exception of local bedrock escarpments, Pleistocene escarpments, and kames.

4.5. HYNDMAN LAKE UPLANDS

Here the glacial drift mantle is patchy and thin, bedrock controls the topography, and the terrain shows relief up to 500 m asl. The surface is described (by Hughes, personal communication) as one showing older weathering in the drift, and it is possible that the conspicuous glacial melt-water channels that occupy the valleys filled today by Hyndman and other lakes, and by the Wolverine River, drained the melt-waters from a glacial maximum marked by the Tutsieta moraine system immediately to the south.

4.6. PEEL PLAIN UPLAND

This is an island-like upland in the Peel Plain Lowlands, to the south of the 'big bend' in the Mackenzie River before it turns north toward the Mackenzie Delta. It rises from the surrounding flat terrain to about 150–350 m and consists of irregular, rolling, hilly terrain with many lakes, mostly quite deep (~10 m), and no bedrock outcrops.

4.7. PEEL PLAIN LOWLANDS

This is an area of extensive, flat, poorly drained lowlands developed on fine-grained alluvial and glaciofluvial silts, with elevations from 20 to 100 m. There are many thermokarst lakes, and there is an extensive development of hummocky peat deposits with permafrost. The area is intersected by the Arctic Red River, and bounded on its east and northwest edges by the Mackenzie and Peel rivers respectively.

4.8. PEEL PLATEAU

This is a region of intermediate topography, between the foothills of the Richardson and Mackenzie Mountains and the Peel Plain Lowlands to the east. It has moderate relief between 200 and 500 m asl. and consists of gently sloping surfaces overlying sedimentary bedrock with variable thickness of glacial drift. Terraces, solifluction lobes, and earth stripes are common on these upper slopes, and level plateaus have many small thermokarst lakes and other permafrost features. The area has several deeply incised river valleys draining the uplands to the west and south.

4.9. THE PORCUPINE PLAINS

This segment of the interior of North Yukon consists of rolling uplands, pediment surfaces, and local lowland basins with poorly drained, fine-grained, and organic deposits, lying in the centre of a large, irregular basin bounded by the mountains and uplands on all sides, except to the west where the Old Crow Flats occur. It is drained westward by the Bell and lower Eagle rivers, which flow into the Porcupine River. The prevailing landscape unit is gently sloping pediment surfaces with local loess deposition and raised beach units, the latter related to the glacial lakes that occupied nearby lowlands. Extensive peat accumulations with polygonal patterns occupy these lowlands.

4.10. EAGLE PLAINS

This large area of low relief (300–600 m) lies south of the Porcupine Plain and straddles the area including and between the upper Eagle River and the upper Porcupine River, where the latter gives way to the Miner River. The predominant feature is extensive, deep, slightly undulating residual deposits developed over the underlying flat shales, siltstones, and sandstones. Loess is present in many areas, and the resulting silty soils with permafrost often support shallow peat deposits on most surfaces. Flat areas are occupied by thicker peats with polygons and palsas.

4.11. OLD CROW FLATS

I have included here both the flat, lowland complex at 300 m asl and the surrounding gently rising pediment surfaces up to ~500 m. The flats themselves occupy ~3,200 km² and consist of thick deposits of unconsolidated clays and silts capped by peat. These are the glaciofluvial deposits laid down during various late Quaternary cycles of glacial lake filling and draining, and finally drained in the early Holocene. The surface has large numbers of thermokarst lakes, the intervening peaty terrain showing extensive ice-wedge polygon development. Many of the larger lakes (1,000–1,500 ha) are rectilinear, oriented in a northwest to southeast direction (Plate 3). All of these thermokarst lakes show active migration and erosion, and few exceed 2 m in depth. The surrounding gently rising pediment surfaces have occasional raised beach features, and hummocky patterned ground is common on many surfaces.

4.12. RICHARDSON MOUNTAINS

This long chain of mountains stretches approximately 300 km north from its southern extremity near the Peel River to within 35 km of the arctic coast (Plate 1). The general configuration of the peaks and ridges reflects directly the complex system of folded, anticlinal structures that make up this unit. Ridges and crests are separated by V-shaped valleys, often with steep mantles of scree and other colluvial deposits. Many summits are rounded and smooth, particularly in areas with soft limestones and shales, while others are capped by scarps and cliffs. The elevations are modest (1,000–1,500 m), but the close proximity of low-lying terrain confers the impression of a montane landscape on the Richardsons. In general the drainage pattern from these mountains reflects the trend of the bedrock faults and folds: north, northeast. Locally, valley glaciers have left lateral moraines that impounded drainage to form lakes; otherwise there are very few lakes in the Richardsons.

4.13. BRITISH AND BARN MOUNTAINS

I have considered these separate tectonic units together from the point of view of physiography and for convenience. The British Mountains are part of the Romanzof Uplift, and they consist of rugged, steep-sided peaks and crests (up to 1,800 m, with relief between 500 and 750 m) showing a general northwest-southeast alignment. They are intersected by the Firth River, and separated from the Barn Mountains to the east by the upper Babbage River. A variety of rocks make up these mountains, including granites, cherts, and quartzites, and the predominant sedimentary shales, limestones, and mudstones. The Barn Mountains are a smaller, lower group of hills rarely exceeding 1,200 m, with relief of about 500 m. To the north the crests are rounded, predominantly limestone cuesta features, with local summits towards the south showing broken, scarp features. The elaborate, *en echelon* sequences of folding result in local steep scarps and screes.

4.14. OLD CROW RANGE

This small montane area is the eastern outlier of the Brooks Range, Alaska. The highest peak, Old Crow Mountain, is 1,270 m high, and the main portion of these uplands consists of sweeping hills about 800–1,000 m high, largely made up of sandstones and shales, but with frequent granitic intrusives, often in the form of large tors with block scree slopes and cliffs.

4.15. KEELE AND NORTH OGILVIE UPLANDS

This large, varied southwest corner of our area is dominated by rounded, predominantly limestone hills 800–1,400 m high, with wide valleys carrying the main rivers (Bluefish River, Salmon Fork River, Fishing Branch River), and numerous tributary streams in roughly v-shaped lateral valleys. Locally, escarpments of limestone occur. South of the Keele Range, in the northern part of the Ogilvie Mountains, limestone is replaced by sandstones and shales as the prevailing surface bedrock type. The boundary between this unit and the adjacent Eagle and Porcupine Plains to the east and northeast cannot be ascertained precisely, as it consists of broad belts of gently sloping, incised pediment surfaces.

3
Vascular Plant Floristics

This short chapter is intended to provide a sketch of the basic data on the vascular plant flora of the area, but will not attempt any description of floristic zones or provinces for reasons that should become clear. The term flora, as used by botanists and particularly by taxonomists and plant geographers, 'has a very definite meaning. It refers to the kinds of plants, taken collectively, which occupy a specific region' (Gleason and Cronquist 1964). So the vascular plant flora of our area of interest consists of roughly 600 species, being an inventory of the vascular plants recorded so far (see Appendix 1).

Plant geographers recognize various categories of flora, or floristic groups, based on the fact that large numbers of species tend to have roughly coincident ranges corresponding to major physiographic divisions of continents. Thus the flora of the North American continent is made up of arctic, boreal, grassland, desert, and other floristic groups, on the grounds that each group has a character- istic species list. Such grouping is largely a matter of opinion, however, and is used primarily as a means of generalizing and communicating in textbooks and other publications.

It would serve no useful purpose to subdivide the flora of this part of northwest Canada into floristic groups because three of its boundaries are completely arbitrary and correspond to no physiographic limits. Also, the number of stations where intensive collecting has been attempted is very few, and it is certain that any serious collecting in the future will add substantially to the preliminary list offered here (Appendix 1).

The compilations of Porsild and Cody (1980) and Hultén (1968) supplemented by collections by Cwynar (1982) and Cooper (personal communication) make it possible to draw up the appended list.

There is some value in assessing the known general geographical affinities of these taxa, to the extent that their ranges are adequately known; Table 2 shows the rough proportions of the main geographical categories to which species areas in northern Canada can be matched with reasonable accuracy. However, it should

Table 2 Approximate percentages of the main floristic elements, as defined in the text, of the vascular plant floras of the study area and of Chukotka (the latter data were abstracted from Yurtsev 1978)

Range type	Study area	Chukotka
Total number of taxa	600	558
Circumpolar	43%	51%
Arctic-montane	45%	47%
Subarctic-boreal	55%	53%
Amphiberingian	19%	25%
Eurasian, absent from North America		24%
American, absent from Asia	38%	
Subarctic boreal	65% of American taxa	
Arctic-montane	35% of American taxa	
Subarctic-boreal	49% of circumpolar taxa	
Arctic-montane	51% of circumpolar taxa	

not be assumed that all the taxa in one category have coincident range boundaries. Far from it – it merely means that there is an approximate tendency for the individual core ranges to conform to the general area, but there are often gaps, overlaps, and other discrepancies. It is probable that if ever a complete set of data points were available to enable the ranges of arctic and boreal species to be mapped in detail, after the broad grouping had been made, as we attempt here, there would be almost as many subgroupings as taxa! One of the weaknesses of floristic plant geography has been an adherence to chorological schemata that are not defensible by objective, analytical methods (see Thaler and Plowright 1973; Harper 1982). If complete data were available, so that it could be established that gaps in ranges were real and not due to lack of field collecting, floristic typologies would serve a useful purpose only if they led to the formulation of testable hypotheses.

In Table 2, circumpolar ranges refer to taxa whose areas of occurrence extend around the northern land masses with or without gaps, regardless of their latitudinal range. Amphiberingian is used in the broad sense of Porsild and Cody (1980) and Yurtsev (1974) to encompass species ranges that straddle the Bering Strait and extend eastwards as far as about longitude 100°W and as far into Eurasia as longitude 100° E. Arctic-alpine ranges refer to species whose ranges are chiefly, but not exclusively, north of the climatically determined lower limits of the arctic and montane areas, and subarctic-boreal are ranges that fall somewhere within the boreal-temperate climatic zones that lie to the south, or altitudinally below these areas.

The table includes for comparative purposes the flora of the area of our immediate interest and the flora of the Chukotka Peninsula.

Obviously these are quite general categories and I have assigned those ranges that do not fit closely quite arbitrarily.

So what can we conclude about the vascular floristic geography of the area? (1) Roughly one-fifth of the flora, just over 100 taxa, is made up of species of amphiberingian range. Some of these are important dominant plants in the tundra vegetation, such as *Salix chamissonis, S. phlebophylla, S. polaris* ssp. *pseudopolaris, S. pulchra, Dryas octopetala, Astragalus umbellatus,* and *Oxytropis nigrescens.* Others are quite localized and minor in terms of their apparent role in the vegetation, although their ranges are continuous (e.g. *Eritrichium aretioides, Boschniakia rossica).* There are many intermediate conditions between these extremes. (2) The rest of the flora is divided roughly equally into a group (~43%) whose ranges are circumpolar, in the broadest sense, and of these about half are arctic-alpine and half subarctic boreal, and a group (~38%) whose areas do not extend beyond North America, of which two-thirds are subarctic-boreal and one-third arctic-alpine. A similar pattern of amphiberingian, circumpolar, and exclusively eurasian is found in Chukotka (Table 2).

Now do these crude statistics lead to any useful hypotheses that are testable? Before we attempt an answer let us sketch the history of plant geography, for it has spawned several hypotheses and much discussion.

The dominant figure in Beringian vascular floristics, the late Professor Dr Eric Hultén, began the study of northern floras with his brilliant, classical doctoral dissertation (Hultén 1937), which stemmed directly from field experience preparatory to his *Flora of Kamtchatka and the Adjacent Islands* (Hultén 1927-8). In the earlier paper he had already signalled the significance of the amphiberingian phenomena in plants (Hultén 1927-8). His thesis was based on two sets of facts: the areal patterns of the arctic-boreal vascular plants occurring in northeast Asia, the Bering Strait area, and adjacent Alaska, Yukon, and the Canadian Arctic; and the geological evidence that large areas of East Siberia and Alaska-Yukon were not covered by ice during the latest glacial period. He demonstrated that the areas of arctic-boreal vascular plants fell into several types, all centred on Beringia, and he proposed that these 'equiformal progressive areas' were the result of survival in ice-free Beringia and subsequent spread east and west into the continents. The resulting patterns were determined by the interaction of the distribution of land and water masses, adaptations to competition and dispersal, and climatic and other environmental stress. The most striking pattern is the amphiberingian, made up on the one hand of plants of Eurasian origin which, he postulated, radiated from centres of dispersal in East Siberia and Amur eastward into North America and, on the other, of species of American origin which survived in Alaska-Yukon and spread into the Old World.

There has been a substantial flow of floristic inventory in the 50 years since Hultén's initial work, notably by Hultén (culminating in his monumental *Flora of Alaska and Neighboring Territories,* 1968, in which a complete bibliographic listing of floras can be found), by Porsild for the Yukon (Porsild 1955), by Anderson (1959) and Wiggins and Thomas (1962) for Alaska, and by the Flora of

USSR (Leningrad). Yurtsev (1963, 1973, 1974) and Yurtsev et al. (1978) have investigated these floristic questions, putting emphasis on the Eurasian flora, and Johnson and Packer (1967) have examined them in the light of cytotaxonomic and evolutionary data.

Although some details have changed, it appears that most taxonomists and phytogeographers accept the central hypothesis of Hultén, that Beringia has served as an important centre of origin of the modern arctic-boreal-montane flora of northeast Asia and northwest North America (Yurtsev 1974; Murray 1981). The role of Beringia as an area of survival is central to the interpretations of the floristic diversity and relatively high frequency of endemism of the area.

Yurtsev (1972) has summarized the large volume of recent (post-1960) Soviet floristic work in northeast Asia, and brought into focus the central facts and problems of Beringian phytogeography. He concludes that the transberingian migrations of plants were analogous to those of animals – predominantly in a west-to-east direction. He suggests that the ancient Angarian flora underwent a profound change in the late Tertiary under the influence of:

(1) the formation of the enormous supercontinent, Eurasia, which stabilized the Asian anticyclone, (2) the formation of a permanent ice cover on the Arctic Ocean as a gigantic refrigerator for the Northern Hemisphere at the end of the Pliocene–beginning of the Pleistocene, which determined the rise of arctic landscapes and flora, concurrently with the extinction of most of the exotic conifers in the northwest ... , (3) the caving-in of inter-montane depressions and the uplifting of the mountain ranges in the northeast in the Late Cenozoic, with a consequent isolation of the interior regions from the primary source of moisture, the Pacific Ocean, but also from the Atlantic cyclones.

Periods of lowered sea level and dry continental climate were characterized by expansion of steppe-tundra forms, centred on Angara, and spreading eastward into northwest America. An east-to-west migration is envisioned, if at all, only for 'the warmer phases (when) Beringia may have been penetrated by American trees such as *Picea glauca*, found in Middle Pleistocene deposits of upper Kolyma.'

Now is this Beringian floristic hypothesis testable? The hypothesis has several parts, and they should be identified before we address the question. The first part is that a core cohort of vascular plants of arctic-boreal affinity but unknown origin *survived* several episodes of glaciation in the Northern Hemisphere in unglaciated terrain in Beringia, and in the intervening non-glacial regimes spread into the continents as they became ice-free. The only method of testing a Pleisto-cene survival hypothesis is to examine palaeobotanical data that show occurrence at known times in the past, and so far only slight progress has been made. Of course many taxa, particularly of upland entomophilous plants, will never be recognized in the Pleistocene record because neither their pollen nor their macro-scopic remains are likely to have been deposited in appropriate sediments. Of the taxa that are represented by their pollen or spores, a large number cannot be identified as to species. A notable exception is *Selaginella sibirica*, an amphiberingian taxon, and it appears to have been recorded continuously on the Chukotka Peninsula during the various undated Pleistocene glacial cycles (Muratova 1973). Otherwise, the taxa that have been identified usually as macrofossils in various Quaternary deposits on both sides of the Bering Strait are of widespread circumpolar, Eurasian, or North American affinity, but none is megaberingian (cf. Colinvaux 1967; Giterman 1980; Katz et al. 1970; Matthews 1974, 1975a; Giterman et al. 1982; Shackleton 1982).

Similarly the hypothesis of *dispersal and migration* of taxa from known or putative centres of survival remains largely untested, mainly because a data base of securely dated palaeofloristic records is lacking. Hopkins et al. (1981) have recognized this fact recently in their speculative but largely undocumented attempt to trace the late Quaternary spread of some North American tree species, both to and from Beringia. We will examine this question in detail in Chapter 6, after the palaeobotanical evidence has been examined.

We can conclude that hypotheses to explain apparent floristic patterns, in this case Beringian, in historical terms involving survival and spread are testable, but they require a full palaeobotanical record. We are very far from a stage when that will be available for northwest America.

We might find it profitable to go back a step and ask: are hypotheses that propose survival during glacial periods the most likely to yield useful, testable conclusions? What other determinants of floristic composition are worthy of consideration? There are at least two, both recognized by Hultén (1937) in his original discussion.

1. *The Modern Physical Environment*

Many examples illustrate this complex of factors. Young (1971) developed his hypothesis that the small size of the flora of St Lawrence Island, in the Bering Sea, was due to 'currently acting ecological, rather than historical factors' into a broad schema for the entire arctic that rests on the assumption that 'the majority of arctic vascular plant species have a northern limit of distribution which is in a state of equilibrium, controlled by the summer temperature regime of the area.' He used this correlative procedure to subdivide the arctic into four zones whose boundaries are determined by sets of selected taxa. The value of this approach is stated to be that, 'on the basis of climatic data, it is possible to make quite accurate predictions of the composition of the flora of an area,' and 'floristic data can be used to indicate the probable summer climatic regime of an area.'

This is an interesting descriptive approach; further, it could lead to experimentally testable hypotheses about the physiological ecology of various species. It does depend, however, both on the accuracy of the assumption of distributional equilibrium and on the accuracy of the boundaries of the ranges of the selected taxa used to determine the 'significant correlations' (used by Young in a purely empirical rather than statistical way) between thermal boundaries and floristic limits. Young makes it clear that his choice of species was arbitrary, and he probably realizes that although the northern limit of, for example, *Equisetum variegatum* coincided 'almost perfectly with the boundary between zones 1 and 2' in 1971, it no longer does in 1982 because of recent collecting in formerly unknown parts of the high arctic (Porsild and Cody 1980). Likewise, although *Melandrium affine* and *Juncus biglumis* were used correctly to typify the northern limits of different floristic types in 1971 (Young 1971, Figure 8), today's data base requires that they be grouped in the same zone (Porsild and Cody 1980). Rowe (1966) has brought into sharp focus one of the central flaws of such floristic groupings used as a basis for ecological or historical generalizations in plant geography; writing: 'Cores of overlapping species ... can be typified in general terms but boundary lines demand attention to particulars.'

As well as climatic discontinuities or boundaries that can be tested as limiting the range of species, the physiography of a region can be used to develop explanatory hypotheses, although these will probably be resolved into ecological factors when testable explanations are developed. For example, many authors (e.g. Hultén 1937, 1968; Yurtsev 1974; Young 1971; Murray 1981) have pointed out that a large fraction of the Beringian endemics are exclusively montane species, and their absence from large arctic areas is due to their failure to migrate through 'lowland areas which are inhospitable to alpine plants ... thus the present distribution of the Beringian plants may depend as much or more on currently acting ecological factors as on historical factors' (Young 1971).

This point of view has been formulated as an attractive and relatively parsimonious development of the Hultén hypothesis by Drury (1969) in a perceptive and overlooked essay. He writes (p 137): 'The thesis [of Hultén] ... might profitably be re-analysed according to habitat complexity in each area suggested to be a center of dispersal. Geographic areas rich in species might first be tested to determine whether or not they are rich in habitat and geographic events leading to speciation.'

2. Intrinsic Biological Factors

A few examples from the enormous number of possible cases will suffice to illustrate that the range of vascular plants can be determined primarily by one or several of their intrinsic reproductive or ecological attributes, functioning independently of the modern physical environment though of course related to its past evolution.

A large number of the taxa whose arctic-subarctic ranges can be described as endemic and / or restricted to small areas are distinctive in their breeding system, or, more generally, their reproductive mode. For example, in the genus *Antennaria*, 26 species occur in arctic-montane parts of North America, and of these at least 14 have very restricted ranges in the high arctic or, in some cases, in the mountains of Alaska-Yukon. It is likely, though not demonstrated in all cases, that these species are polyploids that reproduce apomictically, and that they represent small, disjunct populations given taxonomic rank as a result of 'perennity or by apomixis, both of which tend to ensure population constancy' (Mosquin 1966). An identical situation exists in *Taraxacum*, an apomictic genus for which at least half of the 14 species are of restricted 'endemic' distribution.

Now I have chosen extreme examples to make the point that floristic statistics often subsume such data, and that the real cause is often overlooked: in this case a distinctive genetic system, suggested by many authors as an advantageous evolutionary device (Mosquin 1966; Savile 1972; Löve and Löve 1974), results in small stable populations of species rank whose range cannot be explained by either environmental limiting factors or relict-status historical influences.

Likewise, such biological interactions among species as competition for space, light, nutrients, and other elements of the environment can determine their biological success, and therefore range pattern, and explanations in terms of 'relict status' or historical endemism serve only to obscure the facts. And it is possible that differential migration rates, a phenomenon well documented for Holocene time in both Europe (Birks and Huntley 1983) and North America (Davis 1976), have resulted in species ranges that do not reflect their ecological amplitude.

We should approach interpretive plant geography with these and other caveats firmly in mind. We will examine the specific question of Beringian floristics again in Chapter 6, after the available palaeobotanical evidence has been marshalled and assessed.

Finally, it is of interest to note that the bryophyte flora of our area has been examined in about as much detail as have the vascular plants; Bird et al. (1977) list 263 taxa for a region that includes all of our area plus the Mackenzie Valley and adjacent mountains south to Fort Simpson, Northwest Territories. The vast majority of these taxa are wide-ranging circumpolar (94%), only 3% are endemic to North America, and only 4% (11 taxa) are amphiberingian. These figures contrast with the vascular plant proportions (Table 2) and appear to be somewhat at variance with Crum's (1966) observation that 'Canadian mosses, although broader in their distributions and showing very little endemism, occupy the same general ranges as those ... of flowering plants and ferns.'

Plant geographers still retain that traditional predilection for compiling distribution maps of species ranges, grouping them rather casually, comparing them, usually simply by cartographic inspection, with the limits of various environmental factors, and then developing explanatory generalizations. This process is often interesting for both the author and the reader, but the interest is transitory if the hypotheses cannot be tested by one or other of the only two available methods: the factual data of palaeobotany, or the results of experimental ecology and experimental taxonomy.

4
Modern Vegetation

1. Beringian Vegetation Zones

One of the central themes that pervades discussion of the ecology and origin of floras, faunas, and of course human cultures in the northwest is the affinities with Siberian phenomena. So it is perhaps appropriate to begin a consideration of the modern plant cover of our particular area by examining its relation to the vegetation of adjacent Alaska and the far east of Siberia. This is attempted only at the broad, geographical level by comparing the prevailing complexes of vegetation, or formations, rather than attempting the large task of comparing the areas at the level of individual plant communities.

When we draw up a map at a large scale of the zonal types of vegetation found on the two sides of the Bering Strait (Figure 8), it becomes immediately clear that although the two areas contain similar segments of that circumpolar latitudinal trend or zonation of tundra–forest tundra–taiga (or northern boreal forests), there are major differences in several of the important types of plant cover at the level of floristic composition, but the differences vary widely. For example, coastal lowlands and tracts of flat, poorly drained fine-grained soils in general often support extensive cover of *Eriophorum* (cotton-grass) and *Carex* (sedge) meadows whose structure and floristic composition are remarkably similar on both sides of the Bering Strait. In contrast, the dominant coniferous trees of all upland and lowland boreal woodlands or forests are quite different in the two parts of Beringia: *Larix dahurica* and *Pinus pumila* in Siberia and *Picea glauca and P. mariana* (chiefly) in Alaska-Yukon. These taiga types show many species of low shrub and herb in common, however, usually species of circumpolar distribution, and in fact much of the minor ground cover of these woodlands is uniform over vast areas of the boreal part of the Northern Hemisphere.

In this concise description of the vegetation formations of Beringia I follow Viereck and Little (1972) and Viereck and Dyrness (1980) for Alaska, Rowe (1972) and others for the Yukon, Lavrenko and Sochava (1956) and Aleksan-

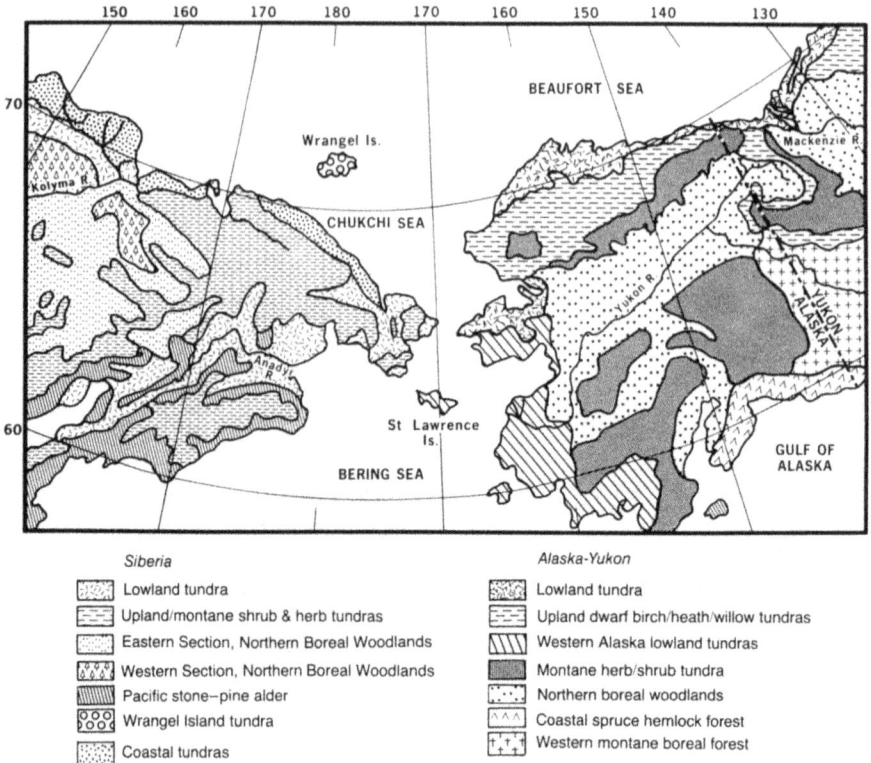

Figure 8 A generalized map of the main vegetation zones of Beringia

drova (1980) for Siberia, and Hare and Ritchie (1972) and Larsen (1980) in general. I will consider Alaska and Yukon together since they have many vegetation types in common. The main vegetation zones are tundra, boreal forests or woodlands, and coastal forests.

1.1. ALASKA-YUKON

(a) *Tundra*
The term tundra is used in North America in its broadest sense, meaning a plant cover dominated by one, or a mixture of the following few elements: lichens, mosses, herbs, and low or prostrate shrubs. At this level of consideration we can identify four main types of tundra.

1. *Lowland tundras*, dominated by members of the Cyperaceae, chiefly *Eriophorum* and *Carex*, forming extensive tracts in the coastal and interior lowland areas. They extend from the northern section of the Seward Peninsula along the entire

coastal lowlands of Alaska, the Yukon, and as far as Nicholson Point, Northwest Territories, on the map of Figure 11 (Chapter 5, Section 2.1).

2. *Heath–dwarf willow / birch tundras* form a complex of structurally rather constant community types, differentiated floristically to some extent, occupying intermediate slopes in northern or high-elevation mountains (e.g. Brooks Range, British Mountains), and moderately drained rolling land in areas north of the arctic treeline (e.g. the Anderson Uplands of the Northwest Territories). The vegetation is dominated by members of the heath family (*Vaccinium*, *Ledum*, *Arctostaphylos*, and others), by low prostrate species of *Salix* (willow), and by the dwarf birch (*Betula glandulosa*).

3. *Montane tundras*. High-elevation mountain habitats, exposed sites on all uplands, and xeric, coarse-grained soils in the arctic zone have a varied plant cover, often quite discontinuous, of tundra communities dominated by three elements whose representation and specific composition vary greatly as a function of such factors as rock type, exposure, and cryoturbation phenomena: herbs, grasses and sedges, and prostrate shrubs. On the general map (Figure 8) no attempt is made to distinguish these montane tundras from the heath-shrub tundras as they usually occupy such discontinuous areas as rocky ridges, screes, and summits.

4. *Coastal tundras of Western Alaska Lowlands*. It is probably unsatisfactory to include in the same group both the Alaska Peninsula Lowlands and the lowland areas adjacent to and south of the Yukon Delta, since the former has distinctive vegetational and floristic elements related more closely to the Aleutian area (Racine 1974), but at this general level of description it hardly merits separation. The area is complex, but the following are the chief tundra types. *Calamagrostis canadensis–Festuca altaica* meadows occupy large tracts of poorly drained lowlands, with *Poa arctica*, *Arctagrostis latifolia*, and *Salix reticulata* on less poorly drained sites (Rausch and Rausch 1968; Hanson 1951). Grass-herb communities of distinctive floristic composition become more prevalent in the Alaska Peninsula Lowlands and at middle elevations (Hanson 1951; Hultén 1962). Stream channels have extensive thickets of *Salix alaxensis*, *S. arbusculoides*, and local *Alnus crispa*.

(b) *Boreal Forests and Woodlands*

There is an extensive literature on the terms used to describe northern vegetation, and I consider it inappropriate here to either extend that discussion or attempt to resolve the rather unimportant questions of nomenclature. The heading of this section refers to all plant cover in this entire area that has an arboreal component, in the form of either continuous stands of trees – usually spruce species, but sometimes tree birch and poplar – or patches of spruce in a tundra matrix, in the form of groves often in climatically favourable sites, or stunted bushes of spruce, usually in tundra near the extreme limit of tree growth. The continuous wooded tracts are often referred to as northern boreal forests, taiga, or boreal woodlands.

However we designate them, they refer to stands of the two spruce species *Picea glauca* and *P. mariana*, sometimes in mixtures, more commonly in stands where one or the other predominates, sometimes associated with tree birch (*Betula papyrifera*), *Populus tremuloides*, or *P. balsamifera*. As the map illustrates, these forests dominate the landscape of Interior Alaska, roughly straddling the immense Yukon valley, extending into lowland areas in the Yukon and into the Mackenzie valley.

The general south to north trend in taiga zonation, so clear in the Mackenzie region and the North Yukon, becomes a partly concentric pattern in interior Alaska because the core of the forested area, the central area occupied by the Yukon River and tributary valleys, is fringed to the north, west, and southwest by mountains. As a result a centre of optimum forest development exists in the interior, near the confluence of the Tanana and Yukon rivers, with gradients of diminishing forest growth radiating north to the Brooks Range, south to the Alaska Range, and west to the Seward Peninsula, correlated with gradients of net radiation and degree-days (Hare and Ritchie 1972; Viereck 1975).

As this is a uniform formation in terms of structure, dynamics, and floristic composition, the detailed descriptions provided later in this chapter for our area of particular interest will complete these brief comments. It remains only to note that the larch, *Larix laricina*, also forms part of this formation, but its distribution is more restricted (to southern areas) than that of the other trees.

At higher elevations in the mountains, and at higher latitudes on the eastern Mackenzie upland plains, the trees become restricted to the least demanding habitats, as permafrost, cryoturbation, lack of winter protection from snow, and shortened growing season play increasing roles in controlling the prevalence of tundra. Some authors (e.g. Aleksandrova 1980) recognize this as a zonal type vegetation formation, and it might be, from the strict viewpoint of a phytogeographer, an appropriate suggestion. However, because of the narrow zone of interaction between Pacific and arctic air masses, and the widespread mountainous terrain, there is not a broad forest-tundra zone in the extreme northwest of North America and there seems to be limited value in recognizing one. Suffice it to add that within the zone mapped as 'boreal woodlands' on Figure 8, the northern and high-elevation fringes, particularly for example on north-facing slopes, can be expected to show open, discontinuous clumps or groves of spruce, rarely *Populus balsamifera*, and the extreme limit of this boundary is spruce krummholz.

(c) *Coastal Spruce-Hemlock Forests*
A coastal fringe of South Alaska and adjacent British Columbia, characterized by a distinctive mild, high-rainfall climate, can be recognized (Figure 8) because of the predominance of several distinctive conifer species, particularly *Picea sitchensis* (Sitka spruce), *Tsuga heterophylla* (western hemlock), with local occurrence of

significant amounts of *Tsuga mertensiana* (mountain hemlock) and *Chamaecyparis nootkatensis* (Alaska-cedar).

(d) *Western Montane Boreal Forests*
Finally in Alaska-Yukon we should note that the western montane section of the boreal forest extends into the west-northwest of the Yukon and a short distance into Alaska along the upper edges of the Yukon River valley. This geographical variant of the boreal forest is distinguished by the partial or complete replacement of the boreal spruces by the high-elevation species *Abies lasiocarpa* (alpine fir), associated at lower elevations with *Pinus contorta* (lodgepole pine), *Betula papyrifera*, and *Populus tremuloides* (aspen).

1.2. SIBERIA

I should preface this section by pointing out that it is based entirely on data from literature as I have done no field-work in East Siberia. The literature is large, and the area is topographically complex. Few authors agree on systems of classifying vegetation and flora; in the interests of consistency, I have followed Gorodkov (in Lavrenko and Sochava 1956), with slight modification in light of accounts of Aleksandrova (1980) and Yurtsev (1972, 1974). It will be clear to informed readers that this account is very general, but more detailed descriptions in English of the tundra zones can be found in Aleksandrova (1980), where references to most of the pertinent literature in Russian are also cited. The basic criterion for defining a unit of vegetation at this scale of mapping is the prevalent cover on zonal, mesic habitats, an approach used generally by botanists.

(a) *Tundra*
We should note that Soviet and North American authors use the terms tundra, polar desert, and steppe in varied ways that have led to some confusion and apparent discrepancies in palaeoecological interpretation. Interested readers can pursue these largely semantic questions in Aleksandrova (1980), Beschel (1970, pp 86–7), and Bliss (1975).
    1. Lowland tundras. This formation is similar both structurally and floristically to the equivalent unit in Alaska / Yukon, and the dominant vegetation is an *Eriophorum vaginatum–Carex* mire complex found on lowland sites throughout the Chukotka Peninsula. Common associated species are *Betula exilis, Salix pulchra, Ledum decumbens, Vaccinium uliginosum, V. vitis-idaea, Hylocomium alaskanum*, and species of *Sphagnum*. This type extends westwards along the broad lowland valley of the Anadyr River, into the valley of the west-flowing Penzhina River, where such species as *Carex soczavaena, C. lugens, Betula middendorffii*, and *B. exilis* are common associates of the dominant *Eriophorum vaginatum*. Other common species, most of circumpolar distribution, are *Vaccinium*

uliginosum, V. vitis-idaea, Ledum decumbens, Empetrum hermaphroditum, and *Arctostaphylos alpina.* With a few exceptions, there is a strong similarity between these lowland mire formations on both sides of the Bering Strait, and both floristic and vegetational differences increase when comparisons are made between upland and montane formations.

2. *Mountain tundra.* This zone occupies (roughly) the 200–1,200 m belt in the complex montane areas of both the Chukotka Peninsula westwards almost to the Kolyma River and the Konyaiskiy-Khribet massifs. Moderate slopes are dominated by *Rhododendron aureum, Phyllodoce aleutica, P. caerulea, Loiseleuria procumbens, Salix arctica,* and *S. chamissonis;* wet meadows are dominated by *Calamagrostis langsdorffii, Aruncus kamtschaticus,* and varied herbs; and rocky, xeric summits have open communities of *Dryas punctata, Minuartia, Selaginella sibirica, Artemisia, Oxytropis,* and *Luzula.*

3. *Coastal tundras.* These form a fringe belt along the northern coasts where the climates are more severe than elsewhere on the East Siberian mainland, and the zonal vegetation is a tundra dominated by *Alopecurus alpinus, Salix polaris, Luzula nivalis, Saxifraga cernua,* and *Eriophorum scheuchzeri.* Poorly drained sites have abundant mosses (*Camptothecium, Drepanocladus, Aulacomnium, Dicranum*) with lichens and sedges.

4. *Moss-lichen tundras.* The extensive, poorly drained lowlands of the area north and west of the Kolyma River are characterized by sedge-moss-lichen tundras with abundant mosses (*Aulacomnium, Drepanocladus, Camptothecium*) and lichens (*Cladonia, Cetraria*).

5. *Wrangel Island tundras.* The vegetation of Wrangel Island is distinctive within Beringia, and most authors describe it as a separate zone or subprovince. Higher elevations have a sparse herb-lichen tundra (polar desert in the terminology of Gorodkov 1958, Lavrenko and Sochava 1956, and others) with species of *Luzula, Oxytropis, Minuartia,* and *Artemisia* and *Salix phlebophylla* associated with abundant lichens and mosses; middle elevations have *Deschampsia, Alopecurus, Salix polaris,* moss-lichen tundras; and lowlands have a sedge-moss tundra complex. Aleksandrova (1980) points out the strong floristic and vegetational resemblance between Wrangel Island and an area near Point Barrow.

(b) *Northern Taiga* (Boreal Woodlands)
The eastern section is a zone of boreal coniferous woodland occupying mesic habitats at middle elevations (300–600 m) in the uplands east and south of the Kolyma River. The dominant tree is *Larix dahurica* (=*L. cajanderi*) with a ground cover dominated by *Empetrum hermaphroditum* (=*nigrum*), *Betula middendorffii,* and various circumpolar mosses and lichens. Lower slopes with impeded drainage have *Alnaster fruticosa* (=*Alnus fruticosa*), *Ledum decumbens,* and *Vaccinium vitis-idaea. Populus suaveolens* and *Chosenia arbutifolia* occur along valley bottoms, and less commonly farther east into Chukotka.

The western section of this formation dominates lower elevations (200–300 m) in the Indigirka lowlands and hills between the Indigirka and Kolyma River valleys. *Larix dahurica* is the dominant tree, and the ground vegetation consists of dwarf shrub cover dominated by *Betula exilis, Vaccinium uliginosum, Empetrum nigrum,* and *Arctostaphylos alpina*.

(c) *Pacific Pine–Alder Scrub Woodlands*
This unusual formation, endemic to East Siberia and Kamchatka, extends from Anadyr along the coastal uplands as far west as the Verkhoyansk Mountains. It forms a pinyon-pine–like scrub on uplands dominated by the 'creeping cedar,' *Pinus pumila,* associated with *Alnus fruticosa, Rhododendron aureum, Betula middendorffii, Rosa acicularis, Ledum decumbens, Vaccinium uliginosum,* and *Boschniakia rossii*.

In summary, we may state that there are strong structural and floristic similarities between the vegetation cover of East Siberia and that of Alaska-Yukon. In general these similarities are greatest for lowland, mire formations and decrease in upland and montane areas. The minor or ground covers of subarctic and boreal woodlands on both continents have many species of shrub, herb, moss, and lichens in common (usually completely or nearly circumpolar species) while the arboreal elements are quite distinct.

1.3. STEPPES AND TUNDRA-STEPPES

Before leaving this general account of Beringian vegetation we should comment on steppe and tundra-steppe vegetation types, mainly because they have, at least in name, played a large role in the discussions of Beringian palaeoecology. Freitag (1977) has offered the useful suggestion that the widespread confusion in palaeoecological literature over the term steppe could be avoided if it were applied only in its strict, original sense, to temperate grassland communities. The confusion is not peculiar to literature on Beringia, but is found in many accounts of the full and late-glacial vegetation of the circum-mediterranean area (Freitag 1977). Tundra-steppe is a term used rarely in North America, but is common in Soviet literature. An important distinction is made by Soviet authors, for example Yurtsev (1974), between true steppes and tundra-steppes. The former are characterized by herb-grass vegetation developed in a continental climate with warm summers, similar to that of the northern prairies or grasslands of the Canadian Western Interior. True steppes do not occur in our area of study, nor in the area of Beringia included in the map (Figure 8). They do form zonal or regional vegetation in areas near and to the east of the Indigirka valley (about 140°30′ E) and farther west in the Yana and upper Lena valleys (between 120° and 130°30′ E and 60° and 68° N). These areas have warm summers (mean July temperature 14° and

19°C; frost-free period 65–125 days; degree-days above 10°C of 1,000–1,500) similar to those of the Northern Great Plains of North America. Steppe-tundra associations are of rare occurrence in the Kolyma and Anadyr watersheds, but they occur in the Chukotka region, scattered in such local, favourable habitats as south-facing riverbanks and terrace slopes. Floristically, these communities do contain a few 'relict' species characteristic of the true steppes (*Festuca lenensis, Helictotrichon krylovii, Carex duriuscula, Thymus oxyodontus,* and *Dracocephalum palmatum*), but many of the dominant and associated plants of these tundra-steppes are found in arctic-subarctic areas on both sides of the Bering Strait: *Selaginella sibirica, Calamagrostis purpurascens, Carex obtusata, C. supina* ssp. *spaniocarpa, Artemisia frigida,* and *Pulsatilla patens.* Plant communities of identical physiognomy and with many species in common with these East Siberian tundra-steppes occur in Alaska-Yukon, often on 'azonal' habitats such as south-facing riverbanks along the Yukon, Porcupine, and Old Crow drainages, or on dry kame summits with gravel soils in the uplands east of the Mackenzie Delta. These tundra-steppe associations also occur on Wrangel Island, and have been the cause of some discussion and controversy in the Soviet literature. We will return to this topic in a later chapter.

2. Present Vegetation of Northwest Canada

The following account of the modern vegetation is based on a survey of published accounts of the plant cover of particular areas as well as my own observations.

There are very few published accounts of the vegetation of North Yukon. Very general descriptions can be found in Rowe (1972), Larsen (1980), Anon. (1974), and Oswald and Senyk (1977). Accounts of the plant cover of particular localities can be found in Drew and Shanks (1965) for the Firth River area in the northwest corner of our region, straddling the international boundary; Johansen (1924), who describes various communities on the arctic coast and on Herschel Island; Lambert (1968, 1972), whose unpublished thesis (1968) remains the most thorough and useful treatment of the vegetation of the North Yukon, dealing with the communities near Canoe Lake in the Richardson Mountains and Trout Lake in the upper Blow River valley; Welsh and Rigby (1971), who provide general notes on the vegetation of plant collecting sites in the British and Barn Mountains, and lowland localities adjacent to these massifs, to both the north and south. Hettinger, Janz, and Wein (1973) provide quantitative analyses of communities from 11 sites in North Yukon extending from the Old Crow Mountains and the upper Rat River to the Barn Mountains and the arctic coastal plain. The unpublished theses and reports of Cwynar (1980), Ovenden (1981), and Ritchie and Cwynar (1976) and the publications by Ritchie (1982) and Cwynar (1982) provide descriptions and analyses of vegetation from several localities, made during the field seasons of the Northern Yukon Research Programme of the University of Toronto.

Accounts of the plant cover of the part of the Northwest Territories included in our area are found in the monograph of Mackay (1963), on the geography of the Mackenzie Delta; this work remains the most authoritative account of the physical geography and vegetation of the area. Two of the most detailed and informative quantitative treatments of the vegetation of particular areas remain unpublished, in thesis form: by Kerfoot (1969) on tundra types on Garry Island, and by Inglis (1975) on the plant cover of the Sitidgi Lake area. Another important thesis, by Gill (1971), has been published in part, and several papers (Gill 1972, 1973a, b, c, 1975) provide useful accounts of the vegetation of the modern Mackenzie Delta. Published descriptions by Corns (1974), Reid (1974), and Ritchie (1974, 1977) contain accounts of the vegetation of the lower Mackenzie River valley and adjacent uplands.

In attempting to draw up a satisfactory framework or grouping of vegetation types, I have referred in detail to the growing body of literature on the vegetation of adjacent Alaska, and I have found helpful the papers of Bliss and Cantlon (1957), Britton (1967), Churchill (1955), Clebsch and Shanks (1968), Gjaerevoll (1954, 1958, 1963, 1967, 1980), Griggs (1934), Hanson (1950, 1951, 1953), Hopkins and Sigafoos (1951), Johnson et al. (1966), Spetzman (1959), Viereck (1966, 1975), Viereck and Little (1972, 1975), Viereck and Dyrness (1980), Webber (1974), Webber and Walker (1975), Wiggins (1951), and Young (1971).

The task of deriving a coherent grouping of plant community types for a large, diverse, poorly described area that is, in addition, very difficult of access is formidable. Ideally one would establish a dense grid of sampling points over the entire area, apply standard methods of tabulation and analysis, and derive a mass of data susceptible to various numerical treatments to produce an objective classification. The possibility that this process will happen here, or in much of Northern Canada, is extremely remote, even if one decided that it would be worth the enormous expense and allocation of personnel. So we proceed by a loose, largely subjective process of culling the existing literature and drawing heavily on personal observations, impressions, and biases.

What I propose below is a schema of vegetation-landscape units that adheres closely in both concept and typology to the groupings proposed by others for similar regions of Alaska (Hanson 1953; Britton 1967; Johnson et al. 1966; Murray and Batten 1977; Viereck 1975; Viereck and Dyrness 1980) and for Northwest Canada (Corns 1974; Inglis 1975; Lambert 1968). I view it as a pragmatic approach that will undoubtedly be changed in the light of further field studies, and I use it simply to facilitate description. Regrettably, there is inadequate information available to include any account of aquatic vegetation types.

The proposed groupings are into three basic physiognomic categories: forest and woodland, tundra, and shrub. The first pertains to vegetation that includes either a continuous, even stand of trees regardless of spacing or a discontinuous stratum of trees, defined by species and not some arbitrary designation of 'tree physiognomy.' Tundra is defined in the usual broad North American way as vegetation dominated by any combinations of low shrubs (chamaephytes in the life-form classification, which is to say shrubs that do not exceed 50 cm in height), herbs, forbs, mosses, and lichens. Shrub refers to communities dominated by woody plants between 50 cm and 2 (3) m in height.

The subdivisions of these three main categories are vegetation-landscape units, which are defined individually and variably in terms of floristic, structural, and mesotopographic (in the sense of Billings 1973) characteristics. It will become clear to the informed reader that there is considerable 'lumping' of categories recognized by some authors, for example Viereck and Dyrness (1980). I make no apology for this, because I have attempted to treat the vegetation only at the level of detail that is appropriate to our main objectives (palaeoecology) and where the available descriptions and analyses permit a reasonably balanced treatment.

These types of landscape, loosely defined in terms of species composition and physical environment, are recognized primarily as tools for communication – for describing the situation to a reader – and for such practical purposes as setting up a utilitarian framework for management or range evaluation. But this approach does not contradict the stochastic or Gleasonian view of plant communities (Gleason 1926; and see useful summaries in McIntosh 1976 and Simberloff 1980). Few ecologists question the notion that plant communities are 'fortuitous abstractions' made up of sets of species, several of which may have overlapping tolerance ranges over considerable segments of environmental gradients.

2.1. FORESTS AND WOODLANDS

(a) *White Spruce Communities on Recent Alluvium*
These are the most productive forests in the area, consisting of closed stands of symmetrical, relatively tall *Picea glauca* (Plate 11). They occur on alluvial deposits in river valleys and deltas, occupying the oldest, uppermost levee of terrace features. These stands are common along all rivers in the boreal, forested section of our area. The northernmost examples of such vegetation in the Yukon are along the Firth River at 68°40′N (Drew and Shanks 1965) and along Dog Creek north of the Old Crow Flats at a similar latitude. In the Northwest Territories there are large tracts of these forests on the levees of the Mackenzie Delta (Gill 1973c) and they extend as far north as 68°50′ in the delta (Mackay 1963) but they occur at 69°N and beyond on the alluvium of the Anderson River. *Picea glauca* may be the sole tree species, or it is sometimes accompanied by *Populus balsamifera*. The minor vegetation of these spruce forests consists of several species of tall (2–4 m) willow (*Salix glauca, S. arbusculoides, S. pulchra, S. alaxensis*), alder (*Alnus crispa* occurs more commonly in northern sites, such as the Mackenzie Delta; *Alnus incana* is scattered there, but occurs as a common element in these communities in the Bluefish, upper Porcupine, upper Eagle, and Peel River watersheds), and, in southern areas, *Cornus stolonifera*. The low shrub and herb

component is dominated by *Arctostaphylos rubra, Pyrola grandiflora, Hedysarum alpinum, Rosa acicularis*, and *Shepherdia canadensis*, while *Hylocomium alaskanum* and *Tomenthypnum nitens* are the dominants of the moss layer. A detailed analysis is found in Appendix 2(a). The soils are poorly developed fibrisols, consisting of 15–35 cm (occasionally up to 50 cm) of fibrous humus overlying undifferentiated, locally gleyed silts, with permafrost at variable depths between 0.5 and 1 m.

(b) *White Spruce on Uplands – Non-Calcareous Parent Material*

These stands of *Picea glauca* occur frequently throughout the area on steeply (20–30°) to moderately steeply sloping surfaces derived from glacial or non-glacial parent materials of non-calcareous origins (Plate 10). I separate these from the next category on the basis of parent material difference because of my consistent observation of significant differences in the species composition of the vegetation on the two types.

The following description is drawn from analyses and accounts of this community type by Inglis (1975) in the vicinity of Sitidgi Lake, by Reid (1974) at many sites in the Mackenzie River valley, by Hettinger et al. (1973) from sites in the Peel River area, and from my own records in the Inuvik, Travaillant Lake, Hyndman Lake, Reindeer Station, and Crossley Lakes areas.

There is considerable variability in the density and age / height / diameter of spruce in these stands, probably due to variations in site characteristics and in fire history. In general the trees are moderately spaced (5,000–8,000 / ha) and they range from 12 to 20 m in height and 25 to 55 cm in diameter at maturity (~200 yr). *Picea glauca* is sometimes the sole tree species, but it is also associated with *Betula papyrifera* and less commonly with *Populus tremuloides* and *P. balsamifera*. These deciduous trees increase in abundance in younger stands with recent fires, and there is a gradient between spruce stands with few deciduous trees to stands of pure birch and poplar that is probably related to fire history. Two of these types are described below in (d) and (e) because they are so widespread in the area.

There is usually a discontinuous shrub layer, ranging from 0.5 to 3 m in height, consisting of *Alnus crispa, Salix glauca, Shepherdia canadensis, Rosa acicularis*, and *Betula glandulosa. Ribes triste* and *Viburnum edule* occur locally. The ground cover varies with local microtopographic changes. Stable sites with mature soils have dense moss-lichen carpets of *Hylocomium alaskanum, Dicranum elongatum, Aulacomnium turgidum*, and *Peltigera aphthosa*, associated with *Vaccinium vitis-idaea, Empetrum nigrum, Ledum decumbens*, and *Vaccinium uliginosum*. Less stable surfaces, or where bedrock is close to the surface and the angle of slope exceeds about 20°, have *Arctostaphylos uva-ursi* and *Juniperus communis* as common shrubs. Complete species lists are given in Appendix 2(b).

The soils of these woodlands are moderately developed brunisols, with a layer of fibrous humus, 5–10 cm thick, overlying a silty-loam layer of 15–30 cm thickness showing some profile development into a reddish illuvial (B) horizon overlying the parent material. In the southern areas of our region, north to about 67°, permafrost rarely occurs above 1 m depth in the soil, but towards the north limit of these stands (approximately 68°50′N) it can be at a depth of 6–80 cm.

Isolated stands of this type have been examined as far north as 69°, in the area immediately south of Eskimo Lakes (Northwest Territories), where they occur in depressions, often roughly south-facing, or flanking esker deposits.

(c) *White Spruce Stands on Uplands – Calcareous Parent Material*

I separate these stands, dominated by *Picea glauca*, from (b) above for the following reasons: the tree layer lacks *Betula papyrifera* and *Populus* species, but includes *Larix laricina* on many surfaces, and the ground vegetation is consistently and markedly different from that of stands in group (b) above. We should note here that I make a similar distinction among upland tundra communities. In both cases this might well be a regional phenomenon – as we noted in Chapter 3, large segments of our area, particularly east of the Richardson Mountains, have parent materials of glacial, Laurentide ice origin with a predominance of siliceous, non-calcareous soils; elsewhere, chiefly but not exclusively west of the Richardsons, calcareous bedrock occurs (limestones, dolomites) in extensive unglaciated areas with the result that derived soils are calcareous.

The descriptions below are based on descriptions of sites in the Firth River area by Drew and Shanks (1965) and on my own analyses from near Inuvik (Ritchie 1977), the South Richardson Mountains (Ritchie 1982), the Bluefish Caves area, and several sites in the Porcupine River uplands and adjacent areas.

There is considerable variation in this type with respect to tree density and size, and in the minor vegetation; no doubt future, more detailed analyses will result in recognition of subsidiary categories. For example I have separated two types on north- and south-facing surfaces in the South Richardson Mountains. However, we are striving for generally applicable groupings here and I do not consider enough data are at hand for the entire area to justify subdivision. We simply recognize that there is great variability, particularly as a function of aspect and slope.

Picea glauca is the dominant tree, associated very rarely on south-facing slopes with isolated, often poorly grown individuals of *Populus tremuloides*, and with *Larix laricina* on generally north-facing surfaces where it may achieve densities of 20–30% that of the spruce. *Larix* reaches its northern limit south and east of that of *Picea* (Zoltai 1973) so it is absent from the northernmost examples of these white spruce woodlands (e.g. those at Firth River and at Bluefish River). Spruce density ranges from maximum values of 6,000 / ha in the Richardson Mountains and the Keele Range to <1,500 / ha on the Campbell Dolomite uplands near Inuvik to 500 / ha in the sparse woodlands on south-facing slopes at the Firth River. Tree size ranges from maximum average stand values at maturity of 8–14 m height and 15–20 cm dbh, to minimum values, at the northern sites, of 5–9 m height and 8–10 cm dbh.

There is a scattered tall shrub stratum in these stands of *Juniperus communis* and *Potentilla fruticosa*, both confined to south-facing slopes. The distinguishing common species in the ground vegetation that are absent from the type (b) stands are *Dryas octopetala, D. integrifolia, Astragalus umbellatus, Carex scirpoidea, Rhododendron lapponicum, Hedysarum alpinum, Tofieldia pusilla,* and *Cassiope tetragona.* These are associated with *Lupinus arcticus, Arctostaphylos rubra, Salix reticulata, S. glauca, Rhytidium rugosum,* and *Tomenthypnum nitens.*

Soils in these communities vary from shallow calcareous regosols with an A horizon of fibrous humus (2–5 cm), and an undifferentiated calcareous silt or sand, to deeper soils of a brunisolic type with a litter horizon (2–5 cm), an A_1 humous-mineral horizon (5–10 cm), and a reddish-grey B horizon overlying the parent material.

In the Keele Range area of northwest Yukon, where gently rolling limestone hills prevail, often with a shallow loess deposit in depressions, this white spruce woodland is replaced on slopes of less than 10° by a mixed *Picea glauca–P. mariana* community, with *Alnus crispa* common in the understorey, and occurrences of *Ledum groenlandicum,* abundant *Vaccinium uliginosum,* and *Betula glandulosa* reflecting increased soil moisture. In these intermediate communities, which I do not distinguish as a separate type at present, the permafrost layer is at about 40–50 cm depth, whereas it is between 50 cm and 1 m for many of the pure white spruce stands.

Detailed species lists are provided in Appendix 2(c) and (d).

Before completing this section on white spruce–dominated communities, I should point out that extensive areas of the rolling uplands in the Hyndman and Anderson uplands, due east of Inuvik, are covered by an open white spruce shrub community. The trees are evenly spaced but sparse (~500 / ha), and the ground cover is a dense community at a height of 50 cm to 1 m of *Betula glandulosa, Salix glauca,* and *Alnus crispa.* There are many fallen, rotting trunks of spruce, indicating, as Black and Bliss (1978) describe, that these areas have been burned and that spruce regeneration has been severely restricted by the dense cover of shrubs. In the same area, spruce (both species) have been eliminated by fire from large tracts, now occupied by scrub and heath with little or no evidence of tree regeneration, giving a false impression of the 'climatic' treeline.

(d) *White Spruce, Birch, Poplar Woodlands on Uplands*
These woodlands have been described by Reid (1974) from several sites in the Lower Mackenzie River and by Hettinger et al. (1973) from sites in the Peel River region. The following account is based on these descriptions and on my own observations in the Inuvik area.

These woodlands consist of variable mixtures of *Picea glauca, Betula papyrifera,* and *Populus tremuloides* on upland, non-calcareous surfaces, and they reach their maximum extent in the southwest extremity of the area, on sandstone-derived soils in the Ogilvie Mountains, in the southern and eastern reaches of the

Eagle Plain, the Peel River region, and throughout the Mackenzie valley and adjacent uplands as far north as about latitude 68°50'.

Tree density is about 7,500 / ha, and ages range from 80 to 100 yr. Fire-scarred trunks and stumps and charcoal horizons indicate that these are secondary communities. Young spruce saplings suggest this successional status, but my observations in the Inuvik area (unpublished) suggest that the presence of small, apparently young spruce can be misleading. In some cases they turn out to be quite old individuals (e.g. 60 yr with a height of ~1 m and dbh of 4–5 cm) whose growth has been suppressed by competition or repeated local, minor solifluction.

Tree heights range from 8 to 12 m for the tall stratum, with dbh 12–18 cm for mature birch and the older spruce trees.

The associated tall shrub layer is dominated by scattered bushes of *Alnus crispa*, while the rather variable ground vegetation has frequent occurrences of *Vaccinium vitis-idaea*, *Linnaea borealis*, *Hylocomium alaskanum*, and *Cladonia mitis*.

The soils are variable, but are basically brunisolic, either eutric or gleisolic. Typically they consist of 1–3 cm of litter and fibrous humus, often with charcoal horizons in the lower levels; a moderately developed red-brown B layer 10–15 cm thick, local B_2 deposition of dark grey clays, and a C layer which is gleyed in poorly drained sites. Permafrost depth varies from 25 to 50 cm.

(e) *White Birch–Poplar (Aspen) on Uplands*

Extensive woodland dominated by *Betula papyrifera* and *Populus tremuloides* occurs on all those upland surfaces in the subarctic region of our area that lack calcareous materials. They are common in the Mackenzie, Peel, Eagle, and Porcupine drainages, but are conspicuously absent in unglaciated areas of the Richardson, Ogilvie, and Keele Range where limestones and dolomites prevail. It is probable that these are secondary, post-fire woodlands, and local regeneration of both white and black spruce can be observed. They occur on uplands, such as river terraces, ground moraine, and kame features, and on residual soils from shale and sandstone bedrock as in the Eagle River canyon area and the Caribou Hills near Reindeer Station.

Tree density varies from 6,000 to 8,000 stems per ha, and tree size from 8 to 10 m height and 8 to 12 cm dbh. The ground cover has scattered *Alnus crispa* and *Salix glauca*, and the minor vegetation includes *Vaccinium vitis-idaea*, *Empetrum nigrum*, *Cladonia mitis*, *Lupinus arcticus*, *Hylocomium alaskanum*, and *Rosa acicularis*. Soils are usually poorly developed incipient brunisols with feeble horizon development.

Types (b) to (e) above can be thought of as stages of forest recovery on upland soils following fire, and it is probable that the scheme developed by Foote (1979) for similar habitats in Alaska applies in our area, since the climate, soils, and species composition are very similar. In the absence of detailed investigations of community succession in relation to fire, I have adopted the scheme here as a practical expedient for descriptive purposes. Foote (1979) describes a shrub willow and deciduous tree sapling stage, 6–25 years after fire, preceded by a short-lived phase of pioneer herbs and mosses (Plate 13). This is followed by birch and aspen stands of variable density that persist for 50–80 years, and, as they open out by natural thinning, show white spruce regeneration from seed (Plate 14). The ensuing mixed white spruce, white birch, and aspen stands (Plate 9) give way to pure white spruce forest of the type (b) above, provided fire does not recur. Many examples of these stages of recovery after fire can be found in the forest zone of our area; the particular species composition, density, and community dynamics vary greatly from the generalized scheme outlined above, according to variation in fire severity and frequency, seed-bed populations, age and composition of the vegetation before the fire, and other factors. Interested readers can find these questions discussed elsewhere; a useful source reference is Viereck and Schandel-meier (1980).

(f) Balsam Poplar Woodlands on Alluvium

Populus balsamifera occurs throughout the area up to the limit of any tree growth, forming pure stands on recent alluvium. Sites have been described by Gill (1973c and elsewhere), Reid (1974), and Hettinger et al. (1973), and I include my own field data from sites in the Bluefish, Porcupine, Old Crow, Eagle, and Mackenzie watersheds in the following summary.

Populus balsamifera is usually the sole dominant of the tree layer, with densities in the range 1,800–2,400 / ha, and heights of 8–12 m (exceptionally to 15 m) and dbh 12–16 cm. There is a varied shrub layer with scattered individuals of *Salix arbusculoides* that reach 5–8 m in height, along with *S. pulchra, S. alaxensis, S. glauca,* and *Alnus.* Towards the south, for example in the Bluefish, Peel, and upper Porcupine watersheds, *A. incana* is the dominant alder in these habitats, but in the Mackenzie Delta *A. crispa* is abundant while *A. incana* is relatively local. The ground vegetation has a small, characteristic set of low shrubs, herbs, and mosses, of which the most common are *Hedysarum alpinum, Equisetum arvense, Arctostaphylos rubra, Rosa acicularis, Mertensia paniculata,* and *Distichium capillaceum.*

The soils of these sites are poorly developed, orthic regosols, consisting of a shallow litter layer (1–2 cm), a zone of humus-stained silt (2–5 cm), and a parent material of fine sands and silts with the permafrost table at depths of 1.0–1.5 m below the surface. These sites are subject to occasional flooding in spring, but less frequently than every year. Occasional seedlings of *Picea glauca* and the consistent presence of white spruce stands or mixed spruce / balsam poplar stands on older levee surfaces confirm the familiar assumption that these poplar stands are an intermediate successional stage to a white spruce forest.

It is pertinent here to note that *Populus balsamifera* in our area, and in the northwest of the continent in general (Murray 1980), occurs beyond the limit of continuous forest, near the extreme limit of any tree growth, usually as apparently clonal stands on local beach soils or alluvium. It appears that the effective seed dispersal and vigorous vegetative growth of this species often result in isolated local patches of trees surrounded by tundra. For example (Plate 20), such an apparently clonal palisade of poplar occurs near Old Man Lake along a shoreline strand, at 69°03′N, 132°27′W.

(g) *Black Spruce Lichen Woodlands on Uplands*
Open lichen woodlands are familiar (at least in textbooks) characteristic features of the taiga, and in parts of the Northern Hemisphere such as northern Quebec they are certainly abundant (Larsen 1980). However, their frequency diminishes sharply to the northwest of boreal America, and in our area they are of quite localized occurrence (Plate 12). The following account draws on information in the works of Inglis (1975) from the Sitidgi Lake area; Reid (1974), Zoltai and Pettapiece (1974), and Hettinger et al. (1973) from several sites in the Mackenzie, Peel, and Porcupine River drainages; and my own observations along the Eagle and Porcupine rivers and near Inuvik.

They consist of rather small stands, usually of *Picea mariana* but occasionally with some *P. glauca* individuals, on rather flat to undulating moraine, terrace, or low pediment surfaces. The trees form an open stratum with densities from 800 to 2,100 / ha. The *P. mariana* trees range from 15 to 20 m in height and 15 to 25 cm in dbh, but occasional to frequent layering produces bushier, many-stemmed clumps. There are occasional trees of *Betula papyrifera* in some sites, and I have recorded highly localized, vegetatively reproducing, stunted individuals of *Populus tremuloides* in these communities in the upper Eagle River area (Plate 12). *Alnus crispa* and *Salix glauca* form a highly discontinuous tall shrub stratum. The dominant lichen element has abundant *Cladonia mitis, C. rangiferina, Cetraria nivalis,* and *Stereocaulon paschale,* associated with *Vaccinium vitis-idaea, Empetrum nigrum, Ledum decumbens, Geocaulon lividum, Vaccinium uliginosum,* and *Betula glandulosa.*

These woodlands are usually developed over soils with moderate to slightly impeded drainage, although I have recorded them also on well-drained sandy gravels. The soils vary from poorly differentiated eutric brunisols, sometimes with slight gleying, brunisolic turbic cryosols in areas with earth hummocks, to dystric brunisols – that is soils with an A_e horizon below the fibrous surface layer indicating mild podsolization. The depth to permafrost is variable, ranging from less than 40 cm to 1 m.

(h) *Black Spruce Open Woodlands with Shrub and Moss*
This is a rather broad category in this treatment, but it is distinguished from the black spruce communities grouped under (i) below by the physiographic position and soil, and by the ground cover. In general, stands grouped under (h) occur on flat to gently inclined surfaces of fine-grained glacial lacustrine, ground moraine,

or loess modified residual soils where the mineral substratum is close to the surface. Earth hummocks are characteristic microtopographic features of these sites.

The following description is based on the reports by Black (1979) from the Inuvik area, Inglis (1975) from near Sitidgi Lake, Reid (1974) and Hettinger et al. (1973) from various sites, and my own observations throughout the Mackenzie region, the Eagle Plains, and the upper Porcupine River.

Picea mariana is the dominant tree, but it forms open stands with densities of 4,000–6,000 / ha, and sizes of 2–7 m height and 4–12 cm dbh. The trees are often layered and frequently assume a crowded branching aspect near the crown. *Alnus crispa* and *Salix glauca* are scattered but constant members of the tall shrub layer, and *Betula glandulosa* is common. The ground vegetation is variable, probably in relation to fire history (Black 1979), but *Vaccinium uliginosum, Ledum palustre, Aulacomnium turgidum, Vaccinium vitis-idaea, Equisetum scirpoidea, Hylocomium alaskanum, Cladonia* species, *Petasites frigidus*, and *Empetrum nigrum* are common. Complete tabulations are found in Appendix 2(f).

These heterogeneous, poorly grown woodlands occupy large tracts of lowland plains (e.g. Eagle Plains) and there is abundant evidence of frequent recent fires. Black and Bliss (1978) have analysed the successional patterns and regeneration ecology of black spruce in these communities, and demonstrate the importance of earth hummocks in providing a microenvironment suitable for seed germination and survival.

(i) *Black Spruce Mire Woodlands*

These wooded bogs or muskegs are found in poorly drained areas where a cap of peat has developed at the soil surface. These communities occur abundantly throughout the tree sector of our area whenever drainage is impeded and stagnant. They often mantle peat palsas and peat plateaus, and they are particularly abundant in the Peel Plain Lowlands. They have been described by Reid (1974), Hettinger et al. (1973), Zoltai and Tarnocai (1974, 1975), and Inglis (1975).

These stunted woodlands have deceptively high tree density values (~4,000 / ha) because the dominant tree, *Picea mariana*, produces dense clumps and thickets with multiple stems by layering. Tree height is 3–5 m, and diameters are variable but always small. The ground vegetation is dominated by low shrubs of which *Betula glandulosa, Salix glauca*, and *Ledum decumbens* are abundant; heaths, chiefly *Vaccinium uliginosum, V. vitis-idaea*, and *Oxycoccus microcarpus*; herbs and forbs such as *Carex lugens* and *Rubus chamaemorus*; and large moss and lichen components of which *Aulacomnium turgidum, Sphagnum balticum, S. fuscum*, and *S. recurvum* are common, as well as *Cladonia rangiferina, C. mitis*, and *Cetraria nivalis* among the lichens. Locally, and particularly in the Peel Plain Lowlands, *Larix laricina* is associated with black spruce, especially where minero-trophic conditions prevail.

The soils are fibrisols, with 0.5–1 m of fibrous moss peat overlying the undifferentiated mineral substratum. Permafrost depth varies from 50 cm in the south of our area to about 30 cm at latitude 68°N.

(j) *Larch Woodlands in Montane Limestone Areas*
Larix laricina has an interesting role in the vegetation of the region. It occurs very locally as a co-dominant of *Picea glauca* on calcareous uplands (type (c) above), and as a co-dominant of *Picea mariana* in minerotrophic bogs and fens in the Peel Plain Lowlands and southwards throughout boreal Canada on such habitats. It becomes locally dominant and also co-dominant with *Picea glauca* at treeline in the Richardson Mountains only on calcareous substrata (Ritchie 1982; Plate 15).

These small stands appear to be confined to steep (25–30°) slopes of northwest aspect, where winter snow accumulates relatively deeply and, because of the aspect, slope angle, and latitude, the relatively limited exposure to direct beam radiation in spring causes the snow to persist. As a result, the soils are locally cool and moist.

The trees are small (5–8 m, maximum 10 m height, 7–12 cm dbh) and densities vary from 200 to 1,000 / ha. The ground vegetation has a scattered tall shrub layer of *Alnus crispa* and *Salix richardsonii*, and the minor vegetation is dominated by *Cassiope tetragona*, *Tomenthypnum nitens*, *Dryas octopetala*, *Arctostaphylos rubra*, and *Festuca altaica*. Detailed floristic tabulations are given in Appendix 2(e).

2.2. TUNDRA

(a) *Salix phlebophylla – Summit Crests and Ridges*
Exposed surfaces in all mountain massifs and on uplands in the northern segment of the area, exclusively with non-calcareous bedrock, usually shales and sandstones, bear a very characteristic, sparse, low tundra dominated by patches of *Salix phlebophylla*, *Arctostaphylos alpina*, *Dryas octopetala*, *Oxytropis nigrescens*, *Betula glandulosa*, *Artemisia arctica*, and *Hierochloe alpina*; the mosses *Rhacomitrium lanuginosum* and *Polytrichastrum alpinum* and lichens *Cetraria nivalis* and *Cornicularia divergens* are common.

Lambert (1968) singles out this community as an association, *Salicetum phlebophyllae*, in his sociological analysis; Cwynar (1980) describes it from North Yukon, and others have identified it as a consistently recurring assemblage on exposed, regosolic, non-calcareous ridges and uplands in Alaska (Spetzman 1959; Churchill 1955; Gjaerevoll 1963; Johnson et al. 1966). Detailed tabular summaries can be found for sites in the North Yukon in Lambert (1968) and Ritchie (1982), and a summary table is provided in Appendix 2(g).

These communities of the most exposed sites grade into the *Betula glandulosa–Ledum decumbens* type described below for middle slopes, and the significant trend appears to be an increase in the proportions of dwarf birch, *Salix arctica*,

Vaccinium vitis-idaea, Empetrum nigrum, and *Vaccinium uliginosum.*

The soils of these communities are poorly developed, thin regosols.

It is interesting to note that this community shows remarkable uniformity in its species composition across mega-Beringia (cf. the Alaskan authorities cited above, as well as Yurtsev [1974], Gorodkov [1958], Lavrenko and Sochava [1956], and others who describe the vegetation of the Chukotka Peninsula).

(b) *Dryas–Carex scirpoidea on Calcareous Summit Crests and Ridges*

Identical surfaces topographically to those of type (a) above, but on calcareous rocks, bear a very different tundra community, and where there happens to be a sharp boundary between calcareous and non-calcareous rocks the contact between the two plant communities is sharply defined (Plate 16). These contrasts are particularly abrupt in unglaciated landscapes where the soil is derived directly from the rock, and no doubt such examples of geological change in space tend to reinforce notions about distinct vegetation 'types.'

These communities of exposed upper slopes and ridges have been described by Drew and Shanks (1965) for the Firth River area and Ritchie (1982) for the South Richardson Mountains. They consist of a closed sward of low vegetation, dominated by *Dryas integrifolia* and *Carex scirpoidea,* associated with *Silene acaulis, Hedysarum mackenzii, Anemone drummondii, Androsace chamaejasme, Oxytropis campestris, Eritrichium aretioides,* and *Kobresia myosuroides.* Floristic details are provided in Appendix 2(h).

On lower slopes where soil depth and winter snow depth increase, these communities show a gradual transition to the *Cassiope tetragona, Salix reticulata, Arctostaphylos rubra* communities grouped below as type (e).

These surfaces show frequent stone stripes and non-sorted circles, and the soil is a regosolic type with little discernible profile development.

(c) *Salix – Upper Slope Snow Patch*

Snow patch, or chionophilous, communities are the floristically distinctive patches of vegetation that occur on uplands where deep, persistent snow accumulates and does not melt completely until mid-summer (roughly the end of July) so that the surfaces are snow-covered for at least 10 months of the year.

Lambert (1968) describes a large number of such communities, and Ritchie (1982) a few from a restricted area, both from the Richardson Mountains. There is considerable variation from site to site, and as it happens all the examples described from North Yukon are from non-calcareous substrates, so it is clear that much more analysis is needed to provide a comprehensive view. Species of *Salix* are often dominants, the most common being *S. reticulata, S. pseudopolaris,* and *S. chamissonis,* and the following are common associates: *Oxyria digyna, Sibbaldia procumbens, Saxifraga punctata, Carex podocarpa, Ranunculus pygmeus, R. escholtzii,* and *Equisetum arvense.*

(d) *Betula-Ledum on Uplands and Middle Montane Slopes*
This dwarf birch–heath tundra, which occurs throughout the northwest and adjacent Alaska on all moderately sloping (5–20°) surfaces where fine materials accumulate to form a soil, where snow does not persist after mid- to late May, and where drainage is not impeded, has been recognized by every botanist who has written on the plant cover of the area. There is wide variation in the species composition, apparently as a function of such factors as soil type (calcareous vs non-calcareous, texture of materials, cryoturbation and related phenomena, aspect, and exposure).

I am presenting here a description of the core type, supplemented by Appendix 2(i) showing the details of its species composition along with that of a few subtypes, drawing from my own records from many sites in the area as well as the detailed analyses of Lambert (1968) and Inglis (1975) and the published synopses by Corns (1974), Reid (1974), and Hettinger et al. (1973).

The vegetation consists of a closed, dense cover of dwarf shrubs whose canopy height varies from 25 to 75 cm, dominated by *Betula glandulosa*, *Ledum decumbens*, *Salix glauca*, *Vaccinium vitis-idaea*, *V. uliginosum*, *Empetrum nigrum*, and *Arctostaphylos alpina* with ground cover of abundant lichens, chiefly *Cladonia* species, *Cetraria cucullata*, and *C. nivalis*, and mosses, *Dicranum elongatum* and *Aulacomnium turgidum* (Plate 19).

Variations on this theme are familiar. For example, at sites with soils derived from calcareous materials *Betula glandulosa* is rare or absent, replaced by *Cassiope tetragona*, *Dryas integrifolia*, *Salix reticulata*, and *Lupinus arcticus*, while *Tomenthypnum nitens* becomes abundant and *Arctostaphylos rubra* is the dominant species of that genus.

Slightly impeded drainage is reflected in a conspicuous *Carex* element in these tundras, often *Carex lugens*, associated with *Eriophorum vaginatum*.

Lower slopes, depressions, and lakeshore embankments where snow duration is longer than average by two to three weeks, either because its winter depth was greater than average or because the surface is somewhat less exposed to direct summer radiation, have a variant of this shrub tundra, dominated by *Cassiope tetragona*. The common associated species are *Betula glandulosa*, *Ledum decumbens*, and both *Vaccinium uliginosum* and *V. vitis-idaea*.

More exposed surfaces with thinner soils, lower soil moisture in summer, and thin snow cover trend towards the *Salix phlebophylla* type, showing significant occurrences of that willow species, along with *Loiseleuria procumbens*, *Carex podocarpa*, *Luzula confusa*, and *Empetrum nigrum*.

Earth hummocks are characteristic of many of these landscapes, particularly on moderate (2–5°) slopes with fine-grained, silty soils, and there is a characteristic association of common species with the microtopographic sites offered by these periglacial features. Inglis (1975), Lambert (1968), and Zoltai and Tarnocai (1974) have described many of these local communities. Hummocks whose summits have not been denuded by instability and erosion are usually capped with a mat of lichens, among which *Cladonia alpestris, C. mitis, C. rangiferina*, and *Cetraria nivalis* predominate. The flanks of hummocks have a varied low shrub community of ericads, *Salix glauca*, and dwarf birch, and the depressions between hummocks where standing water occurs in mid-summer are characterized by *Eriophorum vaginatum, Sphagnum* species, *Rubus chamaemorus, Aulacomnium turgidum*, and *Oxycoccus microcarpus*.

Steeper slopes show a variety of microtopographic features, such as elongated hummocks, stone stripes, and solifluction, and there are usually related minor communities associated with each.

Stable surfaces have regosolic turbic cryosolic soils, consisting of an upper humus layer of 2–5 cm; a horizon of humified organic material incorporated into the mineral soil that varies, usually according to drainage, from 5 to 20 cm in depth; and a B horizon, variably developed, that shows some evidence of illuviation and locally gleying, causing mottling. The depth of the active layer varies from 20 to 50 cm.

(e) *Betula glandulosa–Eriophorum on Lower Slopes*

This type is clearly intermediate between type (d) above and the *Eriophorum* mires of type (f). The only reason I segregate it in this account is that in fact large tracts of the landscape are covered by it. It occurs where drainage is impeded but not completely, on very gentle slopes (4°) and lower montane surfaces, and it certainly grades imperceptibly into the tussock tundra of poorly drained lowlands (type (f) below).

Several authors have provided descriptions of this common type from both the North Yukon and Northwest Territories, and I have drawn particularly on the accounts of Corns (1974), Reid (1973), Lambert (1968), and Cwynar (1980) to supplement my own records.

The dominants of this type are *Betula glandulosa* and *Eriophorum vaginatum*, associated with *Vaccinium uliginosum, Ledum decumbens, Vaccinium vitis-idaea, Carex lugens, Salix planifolia*, and *Petasites frigidus*. A conspicuous moss layer is dominated by *Aulacomnium turgidum, Tomenthypnum nitens*, and *Dicranum muehlenbeckii* (Plate 17).

A very similar association of plants occurs on high-centred ice-wedge polygons, as described below, and it seems clear that a continuum of variation exists along a moisture gradient. Detailed tabular summaries of this type and type (d) can be found in Appendix 2(j).

The soil of this type is gleysolic turbic cryosol, consisting of a variable thickness (10–40 cm) of fibrous humic peat, a number of mineral horizons of mottled (gleyed) silty clay, and an underlying parent material. The active layer rarely exceeds 30 cm. Earth hummocks are common in this type, and the characteristic dense pattern of many exposed, deflated hummock tops gives rise to such terminology as 'spotted tundra' for this landscape unit. The water table is often close to the surface in the depressions between the hummocks.

(f) *Eriophorum vaginatum on Lowlands*

Poorly drained, more or less level lowland plains north of the limit of the woodland zone are covered by a monotonous vegetation dominated by the tussocky cotton-grass *Eriophorum vaginatum*. This type covers many thousands of hectares in the Old Crow Flats and along the entire length of the coastal lowlands to the west and east of the Mackenzie Delta, as well as many more localized sites. It has been described by all of the authors cited above under types (d) and (e) (Plate 18).

The surface of these areas is often marked by fields of low- and high-centred ice-wedge polygons, and the precise vegetation composition is usually related directly to the microtopographic features of these phenomena.

In addition to the dominant cotton-grass, the following are the chief associated species: *Carex lugens, Salix pulchra, Betula glandulosa, Vaccinium uliginosum, Aulacomnium turgidum, Sphagnum rubellum, S. lenense*, and various fruticose lichens. A detailed tabulation can be found in Appendix 2(k).

The soils of these types vary according to the local topography and site history. In general they are cryic fibrisols, characterized by 50–150 cm of peat overlying either organic pond sediments developed on a silt or clay parent material or directly on silts. Permafrost depths seldom exceed 25–30 cm.

(g) *Carex Meadows on Lowlands*

These communities are at the ecological transition between terrestrial and aquatic environments. They occupy low, flat areas adjacent to ponds or thermokarst lake margins, or the parts of coastal lowlands where drainage is poorest. The water table in the thaw season is at or above the soil surface, and these communities often form the cover of the central portion of low-centred polygons. Lambert (1968), Corns (1974), and Reid (1973) provide reasonably detailed accounts of these floristically impoverished assemblages.

The sole dominant is *Carex aquatilis*, associated with variable combinations of *C. bigelowii, C. chordorrhiza, C. rariflora, Salix fuscescens, S. glauca, S. pulchra, Betula glandulosa, Eriophorum angustifolium, Calamagrostis canadensis, Trichophorum caespitosum*, and *Potentilla palustris*, with a surface mat of mosses dominated by *Drepanocladus aduncus, Sphagnum squarrosum, Calliergon cordifolium*, and *Aulacomnium turgidum*. Details are shown in Appendix 2(l).

Very wet low-centred polygons may have ~15 cm standing water, and there *Carex aquatilis* is associated with *Arctophila fulva, Caltha palustris, Hippuris vulgaris*, and other semi-aquatic plants.

The soils of these types fall in the general class of cryic fibrisols, consisting of a variable (40–60 cm) unit of partly decomposed sedge-moss peat overlying either pond or lake organic silts or mineral parent material, usually fine-grained saturated silts. The active layer is 25–40 cm deep.

2.3. SHRUB

(a) *Alnus crispa on Flooded Alluvial Sites*

These dense thickets of alder form part of the distinct zonation that is associated with alluvial surfaces in active deltas and floodplains. These habitats have been described by Gill (1973c). They consist of associations of *Alnus crispa, Salix alaxensis, S. pulchra, S. arbusculoides*, and *S. glauca* on low levees that are flooded annually. The ground cover is discontinuous, with variable composition. The following were recorded from this community type in the valley of the Bluefish River, North Yukon: *Equisetum arvense, Hedysarum alpinum, Mertensia paniculata, Calamagrostis canadensis, Parnassia palustris*, and *Delphinium glaucum*.

The soils are poorly differentiated, immature alluvial silts, and the active layer is 75–150 cm in depth.

(b) *Salix-Dominated Drainage Channels and Stream Margins*

In the treeless zone, drainage channels on moderate slopes, particularly in the coastal plains, intermittent streams, wide stream channels, and stream margins support a reasonably constant set of tall shrubs dominated by *Salix*, sometimes associated with *Alnus crispa*. They often have late melting snow, partly a function of their occurrence in valleys and depressions and partly a result of their effective trapping of the snow. The dominant shrubs are *Salix pulchra, S. planifolia, S. glauca*, and *Betula glandulosa*, and locally *Alnus crispa* is common. The ground vegetation is variable, but the following are common members of this community: *Equisetum arvense, Petasites frigidus, Anemone richardsonii, Polemonium acutiflorum, Valeriana capitata, Drepanocladus aduncus, Hylocomium alaskanum*, and *Drepanocladus uncinatus*.

The soils are immature, gleisolic, and the active layer is at an unknown depth.

2.4. MISCELLANEOUS HABITATS AND ASSOCIATED COMMUNITIES

(a) *Maritime Coasts*

The extensive dune, salt marsh, and related coastal habitats that are characteristic of temperate regions are rare and poorly developed along the arctic coast of our area. Large stretches of the coast consist of a low cliff of supersaturated frozen sediments that thaws out and collapses each summer to the extent of a few metres of horizontal distance (Plate 27). Sea ice, cliff erosion of frozen icy sediments, and low tidal amplitudes prevent the processes of dune and salt marsh development that are familiar at lower latitudes.

However, there are scattered areas, usually associated with river mouths, where small dunes can be found and salt marshes often occur. The following description of the rather simple communities associated with these habitats is based on my observations near the mouth of the Firth and Blow rivers, at

Table 3 A summary of the vegetation–land-form units described in Chapter 4 with equivalent units described from elsewhere in northwest North America

Units described in Chapter 4	Equivalent terms; superscript numbers refer to authors listed below
FORESTS AND WOODLANDS	
(a) White spruce on recent alluvium	*Picea* association[1]; closed evergreen[6, 10]; white spruce–feathermoss[11]
(b) White spruce on non-calcareous uplands	Closed evergreen[6]; white spruce–tall shrub-moss[7]; white spruce–feathermoss[11]
(c) White spruce on calcareous uplands	Spruce woodland terrace[3]; white spruce–*Dryas*–moss[11]
(d) White spruce–birch–poplar on uplands	Open evergreen-deciduous forest[6]; white spruce–birch[5, 11]; closed mixed forest[10]
(e) White birch–poplar on uplands	Open evergreen-deciduous forest[6]; closed deciduous[10]; paper birch–alder–*Calamagrostis*[11]
(f) Balsam poplar on alluvium	*Populus* association[4]; winter deciduous orthophyll[6]; closed deciduous[10]; balsam poplar[11]
(g) Black spruce–lichen on upland	Open evergreen sclerophyll[6]; heath–shrub–lichen (spruce)[7]; open mixed forest[10]; black spruce–feathermoss–*Cladonia*[11]
(h) Black spruce–moss–shrub on lowlands	Open evergreen sclerophyll[6]; spruce–lichen–heath–*Sphagnum*[7]; open mixed forest[10]; black spruce–*Sphagnum*–*Cladonia*[10]
(i) Black spruce mires on lowlands	Open evergreen sclerophyll[6]; black spruce–*Sphagnum*–*Cladonia*[10]
(j) Larch on montane limestone slopes	
TUNDRAS	
(a) *Salix phlebophylla* on summits and ridges	Frost scar[1]; open evergreen microphyllous dwarf shrub[9]; *Salicetum phlebophyllae*[9]
(b) *Dryas–Carex scirpoidea* on calcareous slopes and summits	Frost scar[9]; *Dryas* terrace and alpine tundra[3] alpine sedge–*Dryas*[5]; *Dryas* fellfield[8]
(c) *Salix* on snow-patch surfaces	*Salicetum pseudopolaris / chamissonis / pulchrae*[9]
(d) *Betula–Ledum* on uplands and middle montane	Dwarf shrub heath[1]; low shrub-heath[2]; dwarf shrub types[5]; evergreen orthophyll dwarf steppe[6]; Ericaceous shrub[8]; *Betulo-ledetum decumbentis*[9]; dwarf birch–*Ledum*[11]
(e) *Betula–Eriophorum* on lower slopes	Dwarf shrub heath[1]; birch heath subgroup[2]; *Eriophorum* tussock[8]; *Betulo-Eriophoretum vaginati*[9]
(f) *Eriophorum vaginatum* on lowlands	Sedge cotton-grass heath subgroup[2]; seasonal orthophyll meadow[6]; *Eriophoretosum vaginati*[9]; *Eriophorum* tussock[8]
(g) *Carex* meadows on lowlands	*Carex aquatilis* marsh type[1]; sedge subgroup[2]; sedge meadow terrace[3]; seasonal orthophyll meadow[6]; *Caricetum aquatilis*[9]
SHRUB AND THICKET	
(a) *Alnus crispa–Salix* on flooded alluvial sites	*Alnus crispa* type[1]; tall shrub-herb[2]; *Salix-Alnus* association[4]; alder shrub[5]; willow-alder–*Equisetum*[11]
(b) *Salix* along channels and streams	*Salix* type[1]; *Salix-Equisetum* association[4]; willow shrub[5]; deciduous orthophyll shrub[6]

1 Churchill 1955	5 Hanson 1953	9 Lambert 1968
2 Corns 1974	6 Hettinger et al. 1973	10 Reid 1974
3 Drew and Shanks 1965	7 Inglis 1975	11 Viereck and Dyrness 1980
4 Gill 1973e	8 Johnson et al. 1966	

Clarence Lagoon, Herschel Island, and Shingle Point (all in North Yukon), and on the Tuktoyaktuk Peninsula, and the accounts by Johansen (1924) and Jefferies (1977) from nearby localities.

Sandy beaches within the tidal zone are rarely colonized by more than scattered individual plants, and here *Honckenya peploides* is the only common halophyte to occupy such habitats. *Potentilla egedii* and *Mertensia maritima* do occur occasionally but are more common on stabilized dunes. Dunes are rather rare and small, reaching no more than 1–2 m height and 5–10 width. The chief colonizer is *Elymus arenarius* ssp. *mollis*, which functions to stabilize these sand deposits. The surfaces of stabilized dunes have continuous cover of *Salix pulchra*, *Saxifraga caespitosa*, *S. rivularis*, *Mertensia maritima*, *Chrysanthemum integrifolium*, *Sagina intermedia*, and *Lomatogonium rotatum*.

Saline marshes near these coasts develop on silts and fine sands with high water tables, but the salinity concentrations are rarely high because of sea ice and permafrost melt effects (Jefferies 1977). Salt marsh communities are simple, made up of variable combinations of the following common species: *Carex subspathacea*, *C. glareosa*, *C. ursina*, *Arctophila fulva*, *Puccinellia phryganodes*, *Ranunculus pallasii*, *Potentilla egedii*, *Plantago eriopoda*, *Cochlearia officinalis*, and *Stellaria humifusa*.

(b) *Sandstone and Shale Rubble*

Highly localized deposits of coarse rubble derived from sandstone and shales have an impoverished but characteristic assemblage of plants. For example steep talus slopes of this material in the Richardson Mountains (Ritchie 1982, pp 587–8) support an open woodland of *Picea glauca* and a ground cover of abundant *Dryopteris fragrans*, *Selaginella sibirica*, *Cladonia alpestris*, and *Empetrum nigrum*. Similarly several high strandlines from Glacial Lake Kutchin in the surroundings of the Old Crow Basin have the same community. Spetzman (1959) has described similar communities, but lacking trees, from areas of the arctic slope of Alaska.

5
Vegetation History

1. Tertiary Origins

The published record of late Tertiary plants within our area is small, consisting of palynomorphs from thick sediments underlying the Mackenzie Delta region (Rouse 1977; Norris 1982). Of course a large volume of data exists as a result of the intensive exploratory drilling activities in recent years by several major oil companies, centred on the Mackenzie Delta area and the adjacent Beaufort Sea. However, so far only a small fraction of the palynological results have been released publicly. We will examine these local findings, but we must look also at megafossil and pollen evidence from sites in the western arctic islands and Alaska to gain insight into the history of the plant cover of the area during the past roughly 40 million years. Extrapolation of such data over even quite small distances is a risky procedure so we will take a conservative view in this brief sketch, but the interested reader will be able to pursue the original sources for greater detail.

Formidable methodological problems present themselves when fossils of an age greater than the Quaternary period become the basis for reconstructing past vegetation and environment. In the first place, identification to taxonomic levels below genus becomes impossible, because the taxonomic entity identified rarely has a modern representative. So unless the unlikely though sometimes still applied assumption is made that the ecology of a fossil genus can be based on its modern representative, ecological inference becomes at best speculative, at worst valueless. A second difficulty is that taxonomic categories have been established in the palaeobotanical literature on the basis of single organs, such as leaves, and subsequent re-examination of a wider range of material has shown that there are many errors of identification. These problems have been compounded by the establishment of floristic groupings or elements in fossil assemblages and assigning to them of palaeoclimatic indices on the basis of the modern ranges and tolerances of extant representatives. This insecurely based deterministic view has prevailed among plant geographers, and one of its expressions is the concept of a circumpolar, arcto-Tertiary flora that occupied most of the modern arctic areas in the Palaeocene and Eocene, migrating south *en bloc* during the cooling of the Oligocene, and finally occupying its present areas as the mixed mesophytic forest of parts of warm-temperate Asia and east-north America. Wolfe (1972, 1977) has effectively dispatched the concept in his several analyses of the palaeobotany of northwest America, and it is to his results that we must turn to gain some insight into the probable early origins of the vegetation of our area.

As it happens, the Tertiary rocks of parts of Alaska, particularly sites fringing the Gulf of Alaska, have provided the fullest record of Tertiary fossil plants of any Northern Hemisphere region. Further, it has been possible to correlate and date these beds reasonably securely with marine mollusc assemblages and later with radiometrically dated planktonic foraminiferal zones from cores recovered by various deep sea drilling projects (Berggren 1972; Wolfe 1981). The Tertiary plant fossils of Alaska have been studied for almost 100 years, and the following summary of the extensive findings is based entirely on the publications of Wolfe (1966, 1971, 1972, 1975, 1977, and 1981), Wolfe et al. (1966), and Wolfe and Leopold (1967). These data provide a record of fossil occurrence throughout the Tertiary from the Palaeocene to the lower Pliocene. We will then draw on evidence from a Pliocene site, reported by Hopkins et al. (1971), from the Seward Peninsula of Alaska, and on the results of analysis of the Beaufort Formation on Meighen and other islands in the western Canadian Arctic (Hills and Matthews 1974; Kuc 1974; Hills 1975; Rouse 1977) to provide evidence from early to middle Pliocene time.

In making his reconstructions of Palaeogene vegetation Wolfe (1977) has followed an earlier recommendation by Richards (1955) that analyses of leaf physiognomy, foliar remains of lianas, and patterns of leaf venation are useful indicators of palaeoclimate, based on comparisons with modern analogues. Interested readers are advised to consult Richards's (1955) classical treatise *The Tropical Rain Forest* to gain a conception of the structural characteristics of the several types of vegetation that appear to have prevailed in northwest America during the Tertiary. In modern tropical forests there is a close correlation between climate and the proportion of species with entire margins: Tropical Rain Forests, for example in the Brazil Lowlands, have 88%, Paratropical Forests (e.g. Hainan) 70%, Montane Rain Forests (e.g. New Guinea) 65%, and Subtropical Forests (Taiwan, upper elevations) 45%. Similarly there is a correlation between climate and the relative frequency of woody lianas (highest in the Tropical Rain Forest) and between the branching pattern of leaf veins and climate (high frequencies of main veins with small areoles are common in tropical climates).

Although the earliest indications of a boreal vegetation similar in general aspects to the modern taiga are found in late Tertiary deposits, I will begin this brief account with Eocene records because the changes from that period through-out the ensuing more than 30 million years were part of a global environmental trend that culminated in the Pleistocene glaciations. The locations of the sites

Table 4 A summary of the vegetation reconstructions and inferred climates during the upper Tertiary of Alaska. The chronology is based on Wolfe (1981) and the vegetation and climate on papers by the same author (Wolfe 1969, 1972, 1975, 1977).

MY	Series	Fossil stages	Vegetation	Climate
4	PLIOCENE		Coniferous forest dominated by spruce, pine, with hemlock and fir	Mean annual temperature 2–12°C, coldest month –15° to 0°C. Great variation in the regional climate as a function of ocean proximity and mountain building
8		Clamgulchian		
12	MIOCENE	Homerian	Coniferous forests dominated by Pinaceae with occasional broad-leaved deciduous	
16		Seldovian	North of about 65°N, a region with coniferous forest with a few broad-leaved deciduous genera	Cool moist climate showing great regional variation. Mean annual temperature in the range –5° to 10°, coldest month –10° to 5°C
20				
24	OLIGOCENE	Angoonian	Broad-leaved deciduous forests predominate in the earlier but some reappearance of broad-leaved evergreens at the end of this stage	Slight renewed warming in Late Oligocene. Mean annual temperature 8–12°C, coldest month 0–5°C. Rapid cooling of climate
28				
32				
36		Kummerian	Subtropical, impoverished broad-leaved evergreen forests; evidence for a marine strand forest	Renewed warming to mean annual temperature 13–20°C, coldest month 0–18°C; frost and seasonal rain
40	EOCENE	Upper Ravensian	Broad-leaved deciduous or impoverished subtropical – warm temperate forests, no lianas or buttresses	Cooler, mean annual temperature 11–13°C, coldest month 0°–12° frost and highly seasonal rain fall
44		Middle Ravensian	Marginally paratropical rain forest	Warm, 13–20°C mean annual temperature, 0–18°C coldest month
48		Lower Ravensian	Paratropical rain forest of broad-leaved evergreens with abundant lianas and buttress root structures	Warmest climate, 20–25°C mean annual temperature, coldest month 13–24°C, abundant evenly distributed rain, no frost

referred to below where informative records of Tertiary plants have been found are shown in Figure 9, and the reader should note that the Alaska sites are a selection from a much larger number. The complete details of this impressive record from Alaska can be found in the publications cited throughout this chapter. Table 4 provides a summary of the chronology, vegetation, and inferred environment of the Alaskan evidence described in the following concise narrative.

1.1. ALASKA

(a) *Upper Eocene* (48–37 my)
The Lower Ravensian rocks, of age roughly 48–46 my, are found chiefly in the Prince William Sound region (sites 2 and 4, Figure 9) not far from the modern state capital of Anchorage. They have yielded a diverse flora of genera and, largely on the basis of leaf physiognomic analysis, Wolfe (1977) interprets the assemblage as indicating a Paratropical Rain Forest. He has introduced this term to describe an assemblage that is very similar to that of the Tropical Rain Forest but differs in the reduced percentage of entire-margin taxa (60–75% compared to 80–85% for typical Tropical Rain Forest). The modern structural equivalents of this forest are found in the Hainan Lowlands and Taiwan. The forest was dominated by a two- or three-storeyed canopy of broad-leaved evergreens with abundant lianas and buttress roots. Wolfe (1977) suggests that this rich and complex vegetation grew in the warmest of Alaskan Tertiary climates, with a mean annual temperature of 20–25°C, a mean temperature of the coldest month of 13–24°C, abundant rainfall distributed uniformly throughout the year, and an absence of frost. This contrasts starkly with the modern climate of the Cook Inlet area, where the mean annual temperature is 5°C, one of the higher values for modern Alaska!

Middle Ravensian rocks, aged roughly 45–42 my, are also found in the Prince William Sound area. They have produced a smaller assemblage in terms of numbers of genera and families than the Lower Ravensian. The frequency of entire-margin taxa drops to 54% with a sparse representation of lianas and a pronounced representation of such mixed-mesophytic genera as *Juglans*, *Ptero-carya*, and *Ulmus*. Wolfe (1977) suggests a vegetation that was 'marginally Para-tropical Rain Forest, or more probably in the warmer part of the notophyllous for-est.' (Notophyllous describes a size category of tree leaf intermediate between macrophyllous and mesophyllous.) He suggests a cooler climate, seasonal precipi-tation, a mean annual temperature of 13–20°C, and a temperature of the coldest month of 0–18°C; frosts would have occurred.

The Upper Ravensian is represented by only a single site in the same general area and has too few data for secure reconstructions. Wolfe's tentative (1977, p 33) reconstruction is that the 19 identifiable taxa of megafossil, supplemented by

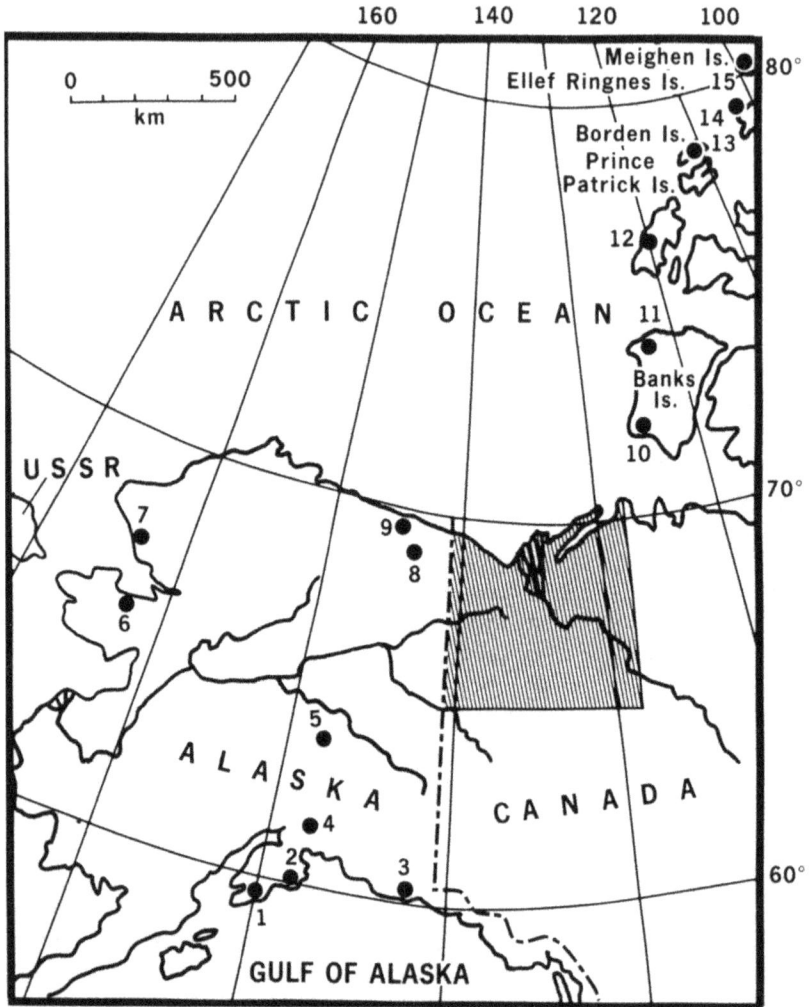

Figure 9 A map of northwest Canada and Alaska showing the locations of the main sites with Tertiary plant assemblages. 1, Seldovia Point on the Kenai Peninsula; 2, Clamgulch; 3, Chugach Mts; 4, Tsadka near Anchorage; 5, Nenana field; 6, Hoogendorn Mine; 7, Kotzebue Sound; 8, Sagavanirktok; 9, Nuwok; 10, 11, Banks Island; 12, Prince Patrick Island; 13, Borden Island; 14, Ellef Ringnes Island; 15, Meighen Island

pollen data, show a marked floristic shift from the Middle Ravensian, with the appearance of such genera as *Salix*, *Juglans*, and *Tilia*, and he suggests a close relationship with broad-leaved deciduous forests of eastern Asia. A climate with a mean annual temperature of 11–13°C, 0–12°C for the coldest months, frost, and seasonal rainfall is suggested.

(b) *Oligocene* (36–23 my)

Kummerian assemblages have been recorded from several localities in the Gulf of Alaska region, chiefly at Nichiwok Mountain just west of Bering Glacier in the southern Chugach Mountains (site 3, Figure 9). These lower Oligocene assemblages indicate a richer, subtropical forest type than from the Upper Ravensian. A leaf margin percentage of 55 with a high frequency of evergreen types is interpreted by Wolfe as an impoverished broad-leaved evergreen forest growing in a climate that had warmed to subtropical conditions with a mean annual temperature of 13–20°C, 0–18°C for the coldest month, frost, and a seasonal precipitation.

The Upper Oligocene is represented by the Angoonian rocks, which have been recorded in Alaska in the Kenai Peninsula (site 1, Figure 9) and at Healy Creek (site 5). The flora is strikingly depauperate, indicating, by both its composition and its leaf physiognomy, a temperate climate. The frequency of entire-margined taxa is low (12%), and Wolfe reconstructs a broad-leaved deciduous forest with a predominance of *Alnus, Betula*, Rosaceae, *Acer*, and Juglandaceae, as well as the gymnosperms *Ginkgo* and *Metasequoia*. Thus there was a rapid cooling of climate in the late Oligocene, beginning about 30 million years ago. However, broad-leaved evergreen genera do occur in the Alaskan late Oligocene, such as *Engelhardtia*, and Wolfe (1977, p 46) offers the hypothesis that at that time 'the temperate vegetation seems to have been enriched, partly through the adaptation of tropical lineages to temperate climates and partly through the diversification of lineages previously present in temperate vegetation.' He continues: 'Families such as Betulaceae, Salicaceae, Rosaceae and Aceraceae, and genera such as *Fagus* and *Castanea*, that were taxonomically depauperate in the Palaeocene and Eocene floras, although present in the Palaeogene subtropical and tropical forests, were able to adapt readily to the new temperate climates and diversified.' The climate of this period probably had mean annual temperatures in the range 8–12°C and 0–5°C for the mean temperature of the coldest month.

(c) *Miocene* (22–6 my)

Seldovian (22–13 my) assemblages are richly preserved in many Alaskan sites, and in general terms those occurring south of 65°N represent broad-leaved deciduous forests of the Mixed Mesophytic Forest type, while those north of 65° are dominated by conifers – e.g. at site 9, Figure 9, by *Picea, Pinus, Tsuga* along with Cupressaceae, *Abies*, and *Larix-Pseudotsuga*, but *Pterocarya* and *Ulmus-Zelkova* also occur. 'These Seldovian assemblages are the richest post-Eocene assemblages in Alaska,' with 'over 100 species of woody plants represented' (Wolfe 1972). Wolfe (1972) suggests that they have developed without any massive migrations *en bloc* as proposed by the deterministic or superorganismic concept formerly espoused by plant geographers under the title 'Arcto-Tertiary Flora,' but rather by geographically variable processes of species elimination, assemblage impoverishment, and enrichment by adaptation of existing taxa and migration of individual entities.

The late Miocene, represented by the Homerian in Alaska, has a fossil flora of roughly 70 woody species, many of which are closely related to living North American taxa. It is interesting to note that the predominant North American endemism of almost all tree species in both the Pinaceae and the broad-leaved deciduous genera has its origin at least as early as the late Miocene. Mid- and north-latitude Alaska appears to have been characterized on lowland sites by this rich coniferous forest with a prominent deciduous element of Betulaceae, Salicaceae, and Ericaceae.

The Clamgulchian rocks in the Cook Inlet region (Wolfe et al. 1966) have been assigned partly to the late Oligocene, partly to the early Pliocene. The assemblage is dominated by conifers, chiefly *Picea*, *Pinus*, *Tsuga*, and *Abies*, and there are sporadic occurrences of *Tilia*, *Ulmus*, *Pterocarya*, and *Diervilla*.

It is of interest to note that the palaeobotany of Tertiary deposits in northeast Asia resembles closely the Alaskan sequence outlined above; interested readers can pursue that topic by referring initially to a review by Biske (1974).

(d) *Pliocene* (6–2 my)
The Lava Camp site (site 6, Figure 9), reported by Hopkins et al. (1971), yielded an interesting assemblage of vascular plants and arthropods, dated by krypton-argon analysis at 5.7 my, and interpreted as 'a rich coniferous forest, dominated by spruce and pine but including hemlock, true fir, either larch or Douglas fir, and one or more members of the Cupressaceae and / or Taxodiaceae.' Similar assemblages have been recorded at Kivalina on the north shore of Kotzebue Sound (site 7, Figure 9), reported by Wolfe (1969). It appears that this coniferous forest vegetation, indicating a climate significantly cooler than that of the late Miocene, represents a stage in the development of the modern North American coniferous boreal forest, which is an even more impoverished version of the quite varied late Miocene precursor to these types.

The late Pliocene is poorly known in Alaska; Wolfe (1972, p 226) offers this summary, partly on the basis of his own unpublished material and partly as a working hypothesis: 'the Pliocene vegetation of Alaska was largely a conifer forest; the climate was apparently cooler than that of the late Miocene, although present evidence is not definite. In the Bering Sea area the late Pliocene saw the disappearance of conifer forest and the concomitant rise of an herbaceous and shrubby vegetation apparently dominated by Cyperaceae, Gramineae, Salicaceae and Rosaceae. Typical tundra plants have not been recorded from this vegetation.' And later he points out that 'there are no known records of lowland tundra vegetation in Alaskan Tertiary rocks, although it is conceivable that on the Arctic slope tundra could have begun developing in the Pliocene. More palaeobotanical data are needed, but I suggest that the development of tundra vegetation is an entirely Late Cenozoic and almost certainly post-Miocene phenomenon.'

1.2. NORTHERN CANADA

Tertiary plant fossils have been recorded from drill core samples taken under the modern Mackenzie Delta and the continental shelf portion of the Beaufort Sea, and from several sites in the modern arctic.

Norris (1982) reports that the spore-pollen flora of middle to late Eocene and early Oligocene sediments of the Richards Formation underlying the Mackenzie Delta contain temperate and warm temperate genera (*Sequoia, Metasequoia, Pinus, Tilia, Ulmus, Castanea, Quercus, Tsuga, Pterocarya*) and a cool temperate element (*Picea, Alnus, Betula, Sphagnum*) that he explains as due to 'long distance transport from cooler perhaps mountainous regions of the northern Cordillera into warm temperate lowlands prevailing in the Mackenzie Delta region.' Also the pollen evidence from these sediments provides a clear demonstration of the mid-Oligocene cooling episode noted above from the Alaskan sites, but also recorded at several sites scattered throughout the globe. The Mackenzie Delta results further corroborate the Alaska data in that they indicate 'an amelioration of climate as warm temperate conditions were re-established' in the late Oligocene (Norris 1982). So far few data are available on Miocene and Pliocene layers sampled in these drilling programs, but Norris (1982) draws on his unpublished results in making the comment that 'Miocene temperate terrestrial floras were replaced by boreal tundra floras' at some time in the Pliocene.

The arctic sites in Canada that have yielded both macrofossil and pollen data are on Banks Island (sites 10 and 11, Figure 9), Prince Patrick Island (site 12), Borden Island (13), Ellef Ringnes Island (14), and Meighen Island (15). The samples come from the Beaufort Formation of Miocene or early Pliocene age, and Hills (1975, and other references cited in this chapter) has made a special study of the palaeobotany of these deposits. The following account is based largely on his papers, and particularly his synthesis published in 1975.

The northernmost site, at Meighen Island, has been investigated by Kuc (1974), who identified approximately 40 taxa of mosses; by Hills and Matthews (1974), who identified 10 vascular plant fossils to species and several to generic level; and by Matthews (1974), who studied the insect fauna, an interesting assemblage of plants was found, including five taxa of conifers. The mosses, as Kuc (1974) concludes, are with one exception all widespread, extant, boreal, or arctic-boreal species with circumpolar distribution. A similar conclusion can be drawn from the short list of partly inconclusively identified vascular plants. It is rather surprising, in light of the evidence, that two of the authors of these reports conclude that the vegetation must have been a forest-tundra. Kuc's (1974) interpretation in terms of a boreal forest appears more concordant with the facts, if we assume that the modern distribution and apparent ecology of these taxa can be related directly to these Miocene or Pliocene representatives. There are taxa that do not occur in the modern forest tundra, but are typical of the boreal forest:

Pleurozium schreberi, Paludella squarrosa, Myrica gale, Pinus, and *Thuja* cf. *occidentalis*. By contrast, none of the species listed has a modern range that is exclusively, or in most cases predominantly, subarctic. In any event, what is readily seen is that there is no evidence of any arctic tundra flora or vegetation at this site.

The Beaufort Formation on Axel Heiberg Island is represented by late Miocene unconsolidated sandstones that have yielded abundant macrofossils of *Picea banksii, Metasequoia distichia, Larix*, and *Alnus* and a single fruit of *Carya*. On Ellef Ringnes Island only pollen evidence has been found so far in this formation, consisting of *Picea, Alnus, Betula, Ledum*, and other ericads. The lower part of the Beaufort beds on northwestern Banks Island is assigned to the early Miocene, and it has yielded a rich pollen record of *Picea, Pinus, Tsuga, Alnus, Betula, Carya, Corylus, Juglans*, and *Tilia*. The uppermost levels show an increase in conifer elements, particularly *Picea banksii, Larix* cf. *omoloica*, and *Tsuga*, and are assumed to be late Miocene in age. Exposures on southwestern Banks Island have produced abundant macrofossils of *Juglans, Picea, Pinus, Larix*, and *Metasequoia*, together with a pollen flora of *Tsuga, Tilia, Carya*, and *Corylus*.

The general conclusion is that the Seldovian to Homerian and finally Clamgulchian changes in Miocene vegetation of northern Alaska were represented by very similar trends on Banks Island. A coniferous forest was established by the late Miocene, and it occurred as far north as 80° latitude in the western arctic.

There is no Pliocene record from the western Canadian arctic-boreal region so one must assume that vegetation similar to the boreal forest assemblages recorded from Alaska existed in the area until the onset of the cooling and then continental glaciation that initiated the Quaternary. During the end of the Tertiary, mountain building was active, and it is assumed (Wolfe 1972) that the appearance of montane habitats was the requisite to the evolution of arctic plants.

Savile (1972) sums up the scant evidence aptly: 'It thus appears that if Meighen Island supported spruce trees until approximately the onset of the Pleistocene, no truly arctic flora can have existed in North America except perhaps on high mountains,' and 'no truly arctic habitats seem to have developed until well into the Tertiary and appreciable arctic plains and marshes have probably existed for little more than 3,000,000 years.' He visualizes the arctic flora as 'a depauperate miscellany from various regions. Not only are arctic genera unknown, but even sections and lesser species groups are scarce.' His conclusion is broadly in agreement with that of other arctic specialists (Tolmatchev and Yurtsev 1970: Löve and Löve 1974; Murray 1981) that arctic vegetation was derived in part from northern boreal forest groupings, from montane habitats, and from plants that occupied exposed sea cliff and beach habitats. Readers are recommended to consult the excellent review paper by Löve and Löve (1974) for a comprehensive assessment of the origins of the arctic and alpine floras as a whole.

We are confronted by a large gap in the record, from the late Miocene until the middle Pleistocene, but we will discover that it is not until the late Pleistocene that continuous, securely dated and thoroughly analysed results have been reported.

2. Early and Middle Pleistocene

Despite intensive research by several large field parties during the past decade, a continuous biostratigraphic record of the early and middle Pleistocene has not been found in northwest Canada. The search for a deposit of continuous primary lake sediment rich in plant microfossils that encompasses the last glacial-interglacial cycle will no doubt continue for several years or decades, but for the present we must acknowledge that the palaeobotanical evidence for a large part of Quaternary time (from roughly 1.8 million to 30,000 years ago) is fragmentary. It comes from four sites, in the collective sense: exposed sections of Bluefish Basin sediments along the Porcupine River; sections of Old Crow Basin sediments exposed along the Old Crow River; a section of Bonnet Plume Basin sediments on Hungry Creek in northeastern Yukon; and miscellaneous peaty strata exposed in coastal and riverbank bluffs along the North Yukon coast and in the adjacent Mackenzie Delta region. We will consider these in turn in the following pages.

2.1. BLUEFISH BASIN

The first thorough palaeobotanical analysis of early and middle Pleistocene sediments from our area remains the most informative and detailed palynological contribution: by Lichti-Federovich (1974), based on laboratory investigation of bag samples collected by O.L. Hughes in 1968 from two riverbank sections at 67°28′N, 139°54′W and 67°31′N, 140°15′W on the Porcupine River between Old Crow village and the Yukon-Alaska boundary (Figure 11). Since this careful study clarification of some aspects of the lithostratigraphy and chronostratigraphy of these sections (Pearce et al. 1982) has made possible the preparation of the summary version of Lichti-Federovich's site 1 pollen diagram with appended chronometric data (Figure 10).

The section can be considered in two parts. The lower half, from river level to about 32 m, consists of fluvial sands, gravels, and silts subdivided by Hughes (1969) into three lithological units numbered 1 to 3 from the river level. Unit 3 has recently yielded a long reversed episode of palaeomagnetism with probable age greater than 700,000 years (Pearce et al. 1982). The pollen assemblages recorded in these levels, and the occurrence of cone fragments of a spruce type intermediate between *Picea glauca* and the extinct, late-Tertiary *P. banksii* described by Hills and Ogilvie (1970), are indirect support for the palaeomagnetic results. The pollen assemblages are quite different from any late Pleistocene, Holocene, or modern spectra recorded from northwest Canada or Alaska. They consist of high frequencies of *Picea*, *Pinus*, and *Betula* associated with consistent smaller percentages of *Corylus*, *Alnus*, and *Salix*. Similar spectra are not found at any of the other early or mid Pleistocene sites in our area. Lichti-Federovich (1974, p 3) draws attention to the interesting and distinctive occurrence of the two *Picea* pollen types, the consistent presence of *Corylus* (hazel) pollen, and the high

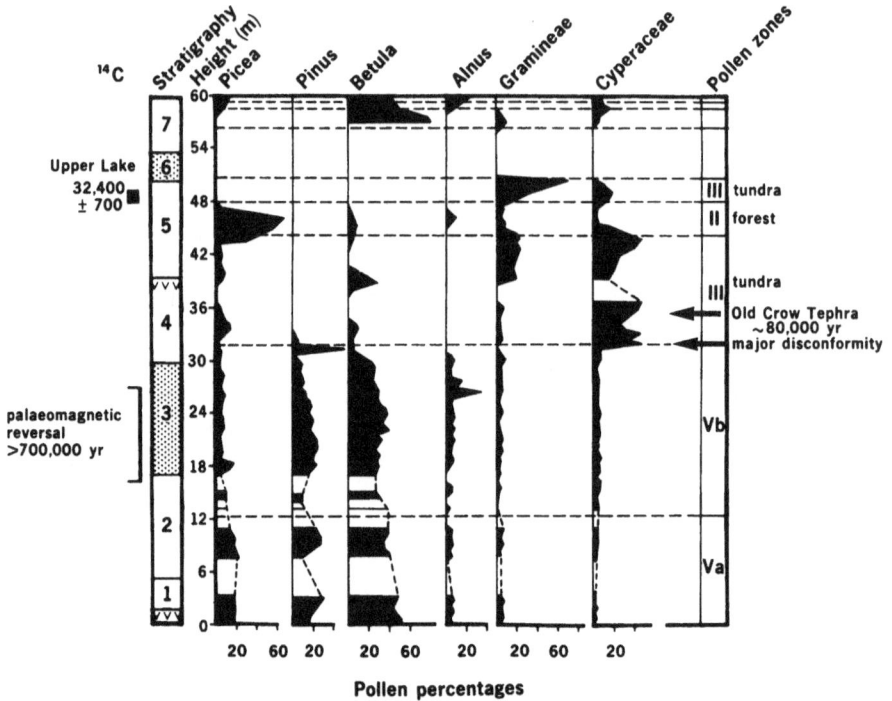

Figure 10 A simplified pollen diagram based on Lichti-Federovich's (1974) analysis and detailed diagrams, incorporating recent data on the Old Crow tephra (Briggs and Westgate 1978) and the reported palaeomagnetic reversal (Pearce et al. 1982)

frequencies (10–40%) of pine, and she draws the tentative conclusion that the vegetation reflected by these assemblages was probably a type of closed boreal forest.

The upper lithostratigraphic unit (unit 3) of this lower portion of the section has been described as a glaciolacustrine deposit (Hughes 1969) and correlated directly with a generally similar unit at sections of Old Crow Basin sediments – the Glacial Lake Old Crow unit – by Jopling et al. (1981). However, Pearce et al. (1982) refute this correlation on the basis of major differences between the pollen assemblages recorded by Lichti-Federovich (1973, 1974) from the two sites.

The upper half of the Porcupine River section consists of about 30 m of sediment subdivided lithologically into units 4, 5 and 6. Unit 4 consists of fluvial layers of silts, sands, and comminuted peats, with abundant evidence of cryoturbation in the upper levels. The pollen evidence consists of predominantly non-arboreal spectra with a relatively rich flora of herb tundra types (*Saxifraga, Phlox, Dryas, Polygonum, Polemonium*, and others), interpreted as representing a tundra

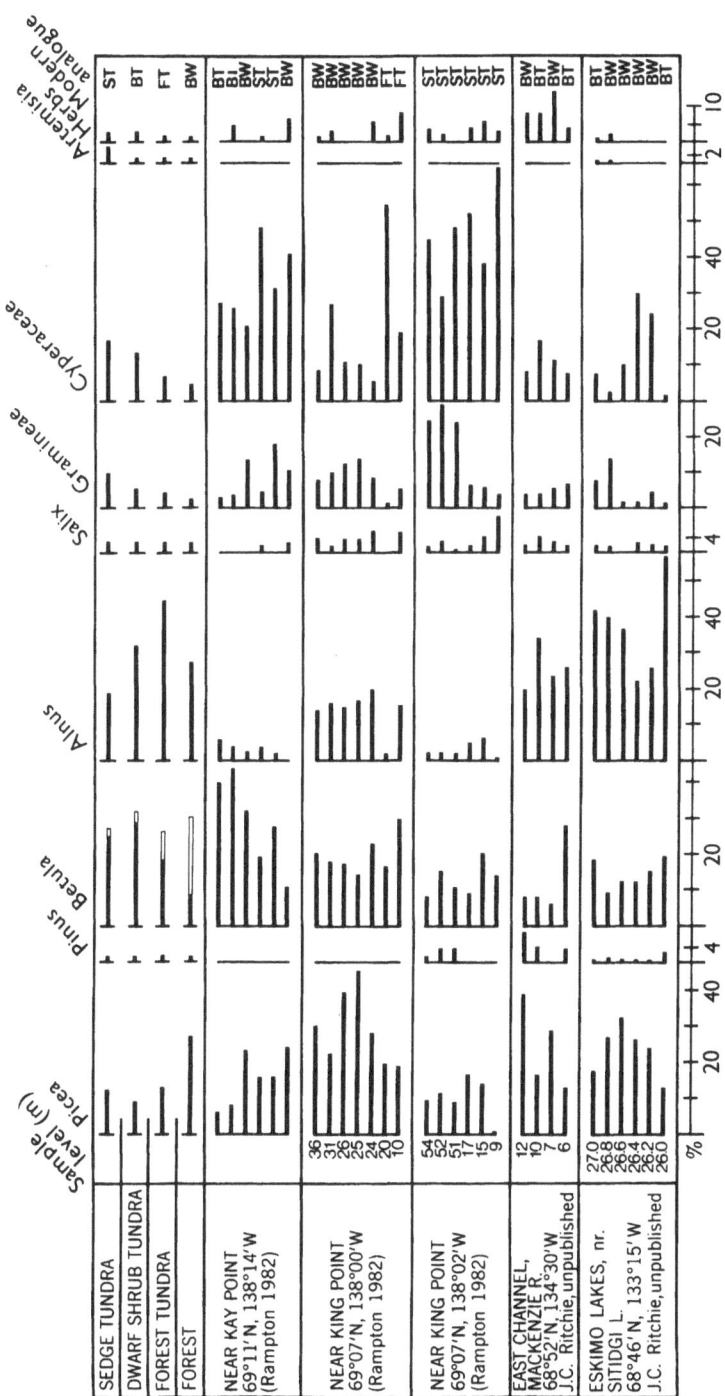

Figure 11 Pollen percentages of the main pollen taxa in samples from five sites with exposures of Pleistocene organic sediments. The top four spectra are mean frequencies of modern samples from the four main vegetation zones indicated, after Ritchie (1974). Open parts of the *Betula* bars refer to tree birch. ST, sedge tundra; BT, dwarf shrub tundra; FT, forest tundra; BW, forest

vegetation and climate. Unit 5 consists of finer sediments indicating pond and overbank depositional environments, with a single 14-C age estimate of 32,400 ± 770 (GSC-952) from shell material, and a fission-track age estimate of a tephra unit in the transition from unit 4 to 5 of approximately 80,000 yr (Briggs and Westgate 1978). The pollen spectra in unit 5 show a clear trend from a shrub-birch tundra assemblage to a spruce-dominated assemblage and a reversion to tundra spectra in the upper samples of the unit. Unit 6 is a sterile deposit of glaciolacustrine clays, assumed to be correlative with the Glacial Lake Kutchin sediments of the Old Crow Basin (Jopling et al. 1981) and the latest stadial maximum of Laurentide ice (Hughes et al. 1981).

2.2. OLD CROW BASIN

The Old Crow River flows east from its Alaskan source and then southwards to its confluence with the Porcupine River (Figure 11). Along its length, and in the banks of some of its tributary streams, many exposures of fluvial and lacustrine sediment occur (Plate 21). They were formed by the complicated series of depositional and erosional processes that occurred in the Old Crow Basin in response to changing drainage conditions. The discovery of great numbers of fossil bones of vertebrates, some apparently modified as human artefacts, focused intensive field research endeavours along this river, particularly the lower reaches. Summary accounts of the history of this flurry of scientific activity can be found in the introductory sections of recent papers by Morlan (1980), Jopling et al. (1981), and Hughes et al. (1981). Samples collected before this period of active team research, by Hughes in 1968, 1969, and 1970, were pollen analysed by Lichti-Federovich (1973). In simplest terms the six exposures sampled consist of a lowest unit of glaciolacustrine silts, referred to as the Lower Lake unit by Hughes (1972) or Glacial Lake Old Crow by Jopling et al. (1981). It has not yielded pollen-bearing samples, and only fragments of plant remains, still unstudied, have been collected (Jopling et al. 1981). The next unit is a thick, variable complex of fluvial and lacustrine beds of sands, silts, and gravels from which Lichti-Federovich (1973) derived pollen stratigraphic results. This unit is overlain by the palynologically barren Upper Lake (Hughes 1972) or Glacial Lake Kutchin unit (Jopling et al. 1981). The middle unit shows variation between exposures in pollen stratigraphy, but Lichti-Federovich (1973) identified a basically tripartite sequence, from the oldest to the youngest layers, of a tundra assemblage dominated at some sites by either dwarf birch and heath, at other sites by herb-grass-sedge pollen types, followed by a *Picea-Betula*-grass-herb assemblage with *Picea* values ranging from 20% to 60%, and finally a return to dominance by non-arboreal pollen types, chiefly grass and sedge (totalling 40–70%) associated with abundant herb pollen types. Lichti-Federovich (1973, p 561) interprets this unit as representing an interstadial sequence of tundra–boreal forest–tundra. Subsequently, the recognition of the Old Crow tephra in the lower levels of the middle unit (as loosely defined above), ice-wedge casts also in the lower levels, and finite 14-C age estimates in the range of 30,000–40,000 yr confirm the interstadial status of these sediments (Morlan 1980) and suggest that they might be correlative with units 4 and 5 (Figure 10) at the Bluefish Basin exposures.

A recent attempt to recover useful pollen records from the Upper Lake unit at the Porcupine River site, employing the effective dispersal and sieving method developed by Cwynar et al. (1979), confirmed Lichti-Federovich's (1974) findings of very low pollen concentrations, poor state of preservation, and an abundance of pre-Quaternary spores suggesting redeposition from older sediments (Ovenden 1981).

2.3. BONNET PLUME BASIN

A roughly 10 m unit of fossil-bearing sands, silts, and clays exposed along Hungry Creek was sampled in detail by O.L. Hughes, R.E. Morlan, and C.E. Schweger in 1976. A collaborative investigation involved analysis of the sediments, radiocarbon and amino acid age estimation, and identification of pollen, bryophytes, vascular plant macrofossils, insects, vertebrates, and archaeological specimens. It resulted in a multi-authored report by Hughes et al. in 1981. The unit is overlain by a Laurentide till, which occurs in many sections throughout the Bonnet Plume area. The location of the Hungry Creek sites studied in detail is shown in Figure 11.

The chronology of the unit is based on a single radiocarbon date of 36,900 ± 300 yr from a portion of a spruce log. The amino acid analysis adds very little to the chronology, merely enabling the authors to speculate that 'an age of between 10,000 and 50,000 years is probably reasonable.' The palaeobotanical data are notable for the rich, thoroughly analysed and informative bryophyte remains. They show conclusively, a result confirmed by the pollen and vascular plant macrofossil data, that the botanical assemblage represents a boreal forest community similar to vegetation occurring farther south today, in the central boreal forest zone of western Canada. The insect analysis suggests in addition the existence of 'numerous dry openings' in the landscape. The authors speculate that these units represent a short, warm interval, 'part of a mid-Wisconsinan interstadial,' but they also point out that the poor chronological control precludes any precise conclusion about the length of time represented by the sediments.

2.4. NORTH YUKON COAST AND ADJACENT MACKENZIE DELTA

Field investigations in 1970, 1972, and 1975 by Dr V.N. Rampton resulted in a comprehensive report on the surficial and Pleistocene geology of the North Yukon coastal plain (Rampton 1982). Samples of organic sediment exposed in coastal and other sections were collected, and several series were analysed for pollen in our laboratory at Scarborough College. The results are communicated in Rampton (1982) and summarized here in diagrammatic form (Figure 11).

Several of these beds lie below a till unit assigned by Rampton to the Buckland glaciation, which he describes as the maximum extension of Laurentide ice along the North Yukon plain. He suggests that this was an early Wisconsinan event and that the underlying pollen-bearing beds are of Sangamon interglacial age or older. The apparent disagreement in the chronology of glaciation between this interpretation and the recent revision by Hughes et al. (1981), which assigns the maximum glaciation to a late Wisconsinan time, makes it impossible to provide a secure time framework for the pollen results. In any event, the data (Figure 11) suggest that regional vegetation types very similar to those of modern spruce boreal forest prevailed on the North Yukon coast during certain episodes, and to those of forest-tundra and sedge tundra during others, based on direct comparison with a set of modern pollen spectra (Ritchie 1974). The modern vegetation is tundra, and treeline lies roughly 100 km to the south, implying that these were episodes of interglacial or interstadial climate warmer than present. All these pollen spectra from the North Yukon are similar to those from units 4 and 5 at the Porcupine River site (Lichti-Federovich 1974, summarized in this volume in Figure 10); they differ from the Hungry Creek spectra (Hughes et al. 1981, p 342) only in their smaller spruce values and higher non-arboreal pollen values. It would not be unreasonable to reconstruct a zonation of boreal forest to open woodland to forest-tundra from the Hungry Creek area to the coast during a past non-glacial episode, assuming that contemporaneity of the strata could be demonstrated. It does seem certain that none of the North Yukon early or mid-Pleistocene pollen assemblages correlates closely with the lower spectra recorded by Lichti-Federovich (1974) from the Porcupine River site, mainly because the large *Pinus* and *Corylus* percentages she recorded there are completely absent from all Yukon coast samples.

In the Mackenzie Delta region two pollen-bearing sites of early or mid-Pleistocene age have been sampled and analysed. Peat beds exposed along the east bank of the East Channel of the Mackenzie River, at 68°52′N, 134°30′W, have been investigated (Terasmae 1959) and referred to by Mackay (1963) as interglacial, on the basis of the pollen evidence for a climate warmer than present, supporting pine and larch, as well as spruce, birch, and alder. The pollen results are included in Figure 11. A similar bed exposed along the south shore of the Eskimo Lakes north of Sitidgi Lake, at 68°46′N, 133°15′W (Plate 23), was described by Mackay (1963, p 17) and sampled by me in 1968. It yielded pollen spectra similar to those of the East Channel samples, indicating the presence of a boreal forest assemblage similar to modern communities found today farther to the south.

The question of whether a pollen assemblage indicating warmer than present conditions at a site should, in the absence of secure chronology, be used to describe an interglacial or an interstadial episode is difficult to resolve; readers are referred to Kukla (1981) for a concise review of several conflicting uses of these terms and concepts.

We can summarize the palaeobotanical record of early and middle Pleistocene easily because it is so sparse and of such poorly known chronology. The history of Pleistocene geological events, while it too has large gaps and apparently conflicting interpretations, suggests that future research might uncover organic deposits that will provide a fuller, ideally continuous record of the long period of Quaternary time that preceded the late Pleistocene.

3. Late Pleistocene and Holocene Record

3.1. TUKTOYAKTUK PENINSULA AND ADJACENT AREAS

We began our palaeoecological investigation here, 16 years ago, not only because it was the area most accessible to the federal research bases at Inuvik and Tuktoyaktuk, but because it is an area of relatively uniform, low-relief topography from which one might expect minimal variability in pollen records between sites. (Readers interested in field and laboratory methods can find a summary in Appendix 3.) In addition the modern vegetation of the area is quite uniform (Corns 1974), with only the distal portion of the peninsula near Atkinson Point showing a distinctive configuration of land-form and vegetation. Further, it seemed a useful place to begin to test ideas about widespread Holocene climatic warming, since the arctic treeline traverses the terrain immediately to the south of the Eskimo Lakes.

Useful pollen, macrofossil, and 14-C results have been published for eight sites from this area (Figure 12), and four additional sites have been successfully cored and are being investigated currently by Ray Spear. Two sites provide continuous, securely radiocarbon-dated primary lake sediment that encompasses the past 13,000 years. I shall discuss these first and then comment on the other sites in the light of the general sequence that the two major sites suggest.

Tuktoyaktuk-5 (T-5) is the informal name given to a small (8 ha) lake at 69°03′N, 133°27′ W (Figure 12) that was cored in 1968 and first reported on in 1971 (Ritchie and Hare 1971). The site is in a broad belt of low, hilly, irregular relief that extends along the southwestern portion of the Tuktoyaktuk Peninsula, north of and roughly parallel to the shore of the Eskimo Lakes. These hills, with relief up to only 60 m, were originally described tentatively as moraines (by Mackay 1963), but more recent opinion, though still not definite, relates these features to 'tentative limits' of the latest glaciation, though 'it is not clear whether [these moraine-like features] are actually moraines or whether thermokarst depressions have developed in these belts to a degree that the landform resembles a moraine' (Fyles, Heginbottom, and Rampton 1972). The most recent summary, in map form (Geological Survey of Canada, Map 32-1979), describes the terrain surrounding Tuktoyaktuk-5 and Sleet Lake as 'rolling hummocky moraine, modified by thermokarst.'

The modern vegetation at this site, and on the entire southwestern two-thirds of the Tuktoyaktuk Peninsula, is quite uniform. The lower parts of moderate slopes (5–10°) have a dense shrub tundra dominated by *Betula glandulosa, Salix glauca, Ledum decumbens, Vaccinium vitis-idaea*, and *V. uliginosum*, with local areas of *Cassiope tetragona. Alnus crispa* is common on these surfaces. Upper slopes and stable ridges have a low shrub-heath dominated by *Betula glandulosa, Salix glauca, Ledum decumbens, Empetrum nigrum, Arctostaphylos alpina*, and *Lupinus arcticus*, while unstable, steep (10–15°) slopes and knolls with coarse sands and gravels support a discontinuous herb cover of *Artemisia frigida, Anemone patens, Potentilla nivea, Lupinus arcticus, Saxifraga tricuspidata*, and *Calamagrostis purpurascens*. Depressions are occupied by *Eriophorum vaginatum–Carex–Betula* meadows on peat. Locally, *Populus balsamifera* clumps, probably clonal, occur along lakeshores (Plate 20) and rare krummholz patches of *Picea glauca* and *P. mariana* have been noted on slopes.

A cored section of 2.5 m of lake sediments yielded a sequence of four clearly distinguished pollen zones spanning the past 13,000 radiocarbon years (Figure 13). The previous report on these data emphasized possible climatic reconstructions (Ritchie and Hare 1971), and we will re-examine that evidence later, but it is now possible to interpret the pollen stratigraphy in terms of possible vegetation history in the light of modern pollen and vegetation data and a greatly expanded network of sites with comparable results.

Pollen assemblage zone T5–1 is separated from zone T5–2 by the *Picea* rise and the associated decreases in relative frequency of *Betula, Shepherdia*, Gramineae, *Artemisia*, and Cyperaceae, and it is characterized by samples with consistent proportions of *Betula* (70–80%, predominantly the dwarf shrub taxon), *Salix* (3–5%), *Shepherdia canadensis* (4–6%), Gramineae (1–4%), *Artemisia* (2–5%), and Cyperaceae (2–4%). There are moderately constant occurrences of *Juniperus*, in low frequencies. Zone T5–1 occupies the radiocarbon age interval from about 13,000 to 9,500 yr.

Assemblage zone T5–2 begins with the rise of spruce above 15% and ends with its decrease again to roughly that frequency at ~5,500 yr. The zone has consistent occurrences of *Picea* (15–50%), *Betula* (40–60%), of which ~50% are of the aboreal type, and *Salix* (2–4%), and low frequencies but constant occurrences of *Myrica*. Pollen zone T5–3 begins at~5,500 BP and is defined by the decrease in spruce to about 15% and birch to about 15% and a marked increase in alder to about 60%. The upper limit of the zone is defined by an increase of birch to 30–40%, a decrease of alder from 60 to 40%, and increases of *Salix*, Ericaceae, Gramineae, and Cyperaceae to about 4% each. These frequencies prevail to the modern sediment surface and characterize zone T5-4.

Sleet Lake (SL) is the second site with a complete record; it will be considered here because its pollen stratigraphy is very similar to that of Tuktoyaktuk-5. It is situated in the Tuktoyaktuk Peninsula, roughly 30 km north-northwest of Tuktoyaktuk-5 (Figure 12), in an area of similar topography and vegetation. It is a small (1 ha) pond, 6 m in depth, that yielded 396 cm of pollen- and macrofossil-rich sediment, sampled and analysed in great detail by Spear (1983, personal

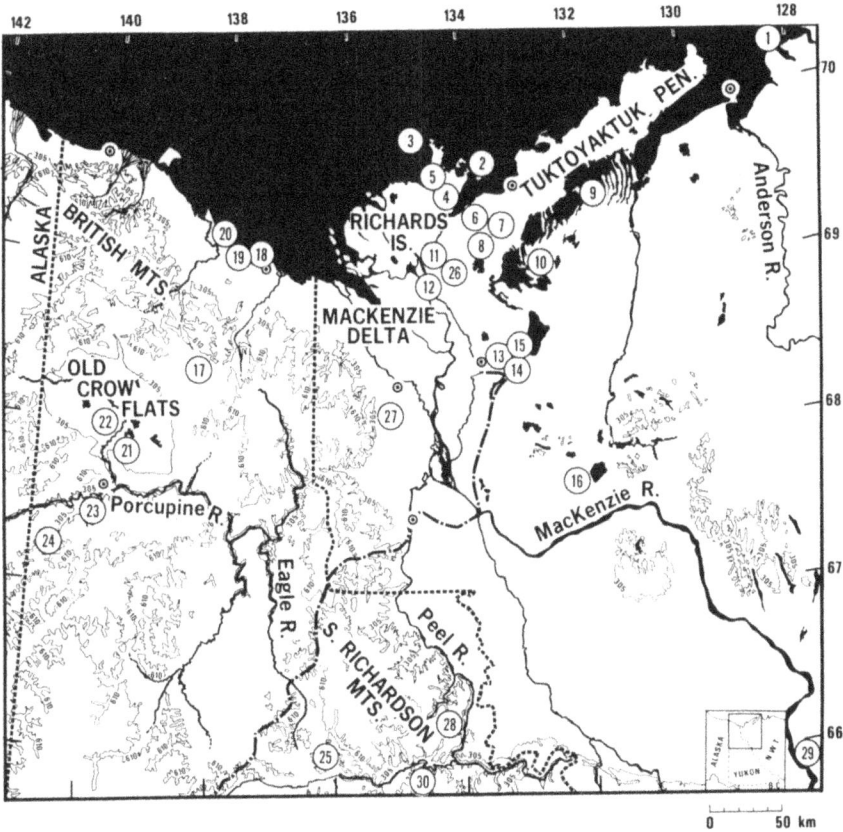

Figure 12 The approximate locations of the sites mentioned in the text. 1, Baillie Islands; 2, Hendrickson Island; 3, Hooper Island; 4, Richards Island pingo; 5, Mayday site; 6, Sleet Lake; 7, Tuktoyaktuk-6; 8, Tuktoyaktuk-5; 9, Eskimo Lake pingo; 10, Eskimo Lake peat bed; 11, Swimming Point; 12, East Channel; 13, Twin Lake Hill; 14, M-Lake; 15, Twin Tamarack Lake; 16, Sweet Little Lake; 17, Hanging Lake; 18, Sabine Point; 19, King Point; 20, Kay Point; 21, Polybog; 22, Old Crow River site complex; 23, Porcupine River site complex; 24, Bluefish Caves; 25, Lateral Pond; 26–29, Delorme et al. (1977) sites 1–4; 30, Hungry Creek (Hughes et al. 1981)

communication); the following comments are based on these published or personal communications.

The lowest pollen assemblage zone, SL-1, spans 12,500 to about 10,000 yr BP, and is identical with zone T5–1 above (Figure 13). Zone SL-2, from 10,000 to ~6,000 yr BP, is very similar to T5-2. It is delimited below and above by increases and decreases respectively of *Picea* percentages from less than 2% to 31%, and from more than 20% throughout the zone to less than 10%. *Betula* frequencies drop abruptly at the boundary between SL-2 and SL-3. *Myrica* percentages are higher in SL-2 than in T5–2. There is a similar separation into two assemblage zones 4 and 5 in the upper samples (6,000 to 0 yr) at both Sleet Lake and T5. *Betula* increases, *Alnus* decreases, and Ericales, Gramineae, and Cyperaceae all increase to their modern values, which are very similar at both sites.

Pollen influx data from this site (Figure 14, after Spear 1983) confirm the percentage diagram, but show a more accurate registration of the spruce and alder changes. In addition, Spear's (1983) preliminary analysis of spruce pollen size, using the method developed by Birks and Peglar (1980), indicates that white spruce (*P. glauca*) arrived first and dominated the early Holocene forests while black spruce (*P. mariana*) became abundant about 7,000–8,000 yr ago.

Pollen analysis of several 14-C dated but incomplete sections of organic sediment from this region have been published, and they corroborate in general the sequence established at the above two sites.

Hyvärinen and Ritchie (1975) presented relative and influx data from samples taken from exposed, more or less vertical faces of two eroded pingos – one on Hendrickson Island at 69°32′N, 133°34′W, between Tuktoyaktuk and Richards Island, the other at 69°24′N, 131°40′W, on the south shore of Eskimo Lakes, 53 km east of Tuktoyaktuk. Both diagrams (Hyvärinen and Ritchie 1975, Figures 3, 5, and 6) show the lowest birch-dominated assemblage, replaced at roughly 9,000 yr BP by the spruce-birch zone, which is in turn replaced at roughly 6,000 yr BP by a zone dominated by alder and birch with reduced frequencies of spruce. The locations of these sites are shown in Figure 12.

Another ruptured pingo deposit, on Richards Island at 69°26′ N, 134°30′W (Figure 12), was sampled in 1966 by J.G. Fyles, and the results of pollen analysis and a single 14-C determination were published (Ritchie 1972). The lowest levels show the same dwarf birch dominated spectra as the above sites, and the spruce-birch zone begins at about 9,390 yr BP, delimited by the sharp increase in *Picea* pollen frequency. There is a marked alder rise in samples from mid-section levels, but only the spruce rise was 14-C dated.

In the same summer Fyles sampled an exposure of pond sediments, partially eroded by coastal collapse of a cliff on the north shore of Hooper Island, at 69°42′N, 134°52′W (Figure 12). A pollen diagram from the widely spaced samples (Ritchie 1972) shows a very similar sequence to that of the above site. A single radiocarbon estimate of 11,800 yr BP was obtained from material within the birch zone.

An undated core of sediment from a lake 11 km north-northeast of Tuktoyaktuk-5 was sampled at wide intervals and produced a pollen diagram with the same four pollen assemblages as at the Tuktoyaktuk-5 and Sleet Lake sites (Ritchie 1972, Figure 5).

Finally, an exposed peat bed on the north shore of Eskimo Lakes, at 69°15′N, 132°20′W, was sampled by J. Ross Mackay in 1961 and pollen analysed by J. Terasmae. Their results (Mackay and Terasmae 1963) suggest that the site is truncated as the lowest organic sediments gave a radiocarbon age of 8,000 yr BP and the pollen stratigraphy is not clearly differentiated. The authors did not

Table 5 Spruce macrofossils from the Tuktoyaktuk Peninsula (after Spear 1983)

Locality and reference	Material and radiocarbon age
69°07', 133°16'; Ritchie and Hare (1971)	stump, 4,940 ± 140, GSC-1265
69°28', 132°35'; Burden et al. (1975, personal communication)	twigs, 5,945 ± 100.
69°27', 132°10'; Spear (1983)	stump, 6,620 ± 70, GSC-3239
69°16', 132°20'; Spear (1983)	log, 6,380 ± 70, GSC-3242
69°16', 132°20', Delorme et al. (1977)	wood, cones, 5,700 ± 100 BGS-215

propose any zonation of the diagram, but there is a general agreement with the stratigraphy described above for mid to late Holocene parts of sections at other sites in the area.

The pollen records from the above seven sites are remarkably similar, and it is closely correlated sequences of this type that confirm the great value and reliability of pollen analysis of lake sediment as a biostratigraphic tool. The three main biostratigraphic units are the birch–willow–soap-berry from 13,000 to 9,500 yr BP, the spruce unit from 9,500 to 6,000 yr BP, and the birch-alder-ericad from 6,000 to the present.

The pollen record has been supplemented by discoveries of macrofossils of spruce on the peninsula, and these data greatly strengthen the vegetation reconstructions. Spear (1983) has collated these data in tabular form (Table 5), which shows that well-preserved radiocarbon-dated twigs, cones, and stumps of *Picea* have been found from five sites between 69°07'N and 69°28'N, all between 132° and 134°W (Figure 12). These finds fall within the radiocarbon age span 5,000–6,600 BP. Spruce seeds and needles occur frequently in the Sleet Lake sediments, between 9,800 and 5,900 BP.

Vegetation Reconstruction
The question now arises: what vegetation patterns can be reconstructed from these pollen data? A survey of modern pollen deposition in this area (Ritchie 1974) showed that tundra and boreal woodland sites can be characterized easily in terms of pollen spectra, but the forest-tundra ecotone is less clearly separable. Direct inspection of these modern assemblages and the Sleet Lake and T-5 diagrams indicates that reasonable matches can be found between all except one of the sub-fossil zonal spectra and modern frequencies. The exception is the oldest zone, dominated by pollen of birch (80–95%), willow, and soap-berry. In an effort to improve the procedure of reconstruction, I have calculated R_{rel}-values for the Tuktoyaktuk Peninsula area, using the percentage cover data for the vegetation of the area (Tables I and II in Corns 1974) and the percentage pollen data for modern lake sediment samples from the same area (Ritchie 1974). The results are shown in Table 6, and recalculated or corrected percentages of the main pollen

Table 6 R_{rel} values for the main pollen taxa calculated from available modern pollen data in lake sediments (Ritchie 1974) and quantitative vegetation
data from the Tuktoyaktuk Peninsula in Corns
(1974) and from the Inuvik area in Ritchie (1974).
The number of vegetation sample sites is small so
these values should be considered as preliminary
estimates.

Pollen taxon	Tuktoyaktuk area	Inuvik area
Picea	absent	0.5
Larix	absent	0.2
Populus	absent	0.6
Betula	4.4	3.0
Alnus	13.4	11.8
Salix	1.7	0.6
Juniperus	1.0	1.0
Ericaceae	0.3	0.1
Gramineae	5.5	3.4
Cyperaceae	3.7	2.2
Artemisia	10.6	5.0
Shepherdia	1.9	absent

types are shown in diagrammatic form in Figure 15. The limitations and advantages of *R*-values are well known and have been fully discussed recently by Birks
and Birks (1980, p 219 *et seq.*) and Parson and Prentice (1981).

The main effects of correcting the Sleet and T-5 diagrams are to depress the
birch and alder frequencies and to increase the proportion of ericad and *Shepherdia*. The fact remains, of course, that a modern equivalent of the zone 1 spectra
is absent. I suggest that immediately following deglaciation, at about 13,000 BP,
the landscape was colonized by a dense shrub tundra on upland sites, with *Betula
glandulosa* dominating on most middle slopes and low ridges and *Shepherdia
canadensis* and *Juniperus communis* dominating steeper, particularly south-
facing, slopes. Ridge summits and kame-like mounds, usually with sand-gravel
soils, supported an open sedge-herb-grass community, probably dominated by
Artemisia, Calamagrostis canadensis, Anemone species, legumes such as *Hedysarum* and *Lupinus*, and *Carex* species. Lowland, poorly drained sites between
ridges and in depressions supported a *Salix* community with grass-sedge marshes
in the wettest sites. Spruce migrated into the peninsula from some southern
and / or western refugium (to be discussed later) and while the pollen began to
arrive about 10.5 14-C yr ago actual tree movement onto the landscape probably
occurred a few centuries later. Spruce percentages, in the uncorrected diagrams
(Figure 13) reached 35–40% by about 7,000 yr BP at T-5 and 30% by the same time
at Sleet Lake. These increases were accompanied by increases of ericad pollen at

both sites, and the corrected diagrams (Figure 15) show that heath family species made up about 20% of the landscape at the time while dwarf birch had declined to about 20%. Measurements of annual ring widths of two of the dated fossil spruce stumps show that their mean radial growth rates were roughly equal to those of modern white spruce on intermediate drainage (i.e. mesic) sites in the main boreal forest at latitudes in the central Mackenzie Valley. One possible inference from this small sample of tree ring evidence and from the pollen frequencies is that from about 9,000 to 5,000 yr BP the Tuktoyaktuk Peninsula was covered by continuous spruce forest on all sites except those with extreme conditions of drainage or stability. Xeric sites, such as gravel-sands on ridge summits, probably supported local herb tundras or willow shrub communities and wet lowland sites probably were occupied by treed mires dominated by *Picea mariana* and *Sphagnum*. The pollen stratigraphy of the eroded pingos and other sites on the Tuktoyaktuk Peninsula suggests that the northern limit of this spruce woodland zone was roughly at 69°50′ N, although the precise details of the extent and timing of this episode will not be shown until current analyses by R. Spear (personal communication) have been completed.

Later, between 8,000 and 6,000 yr BP *Picea mariana*, *Larix laricina*, *Betula papyrifera*, and finally the two *Alnus* species spread into the area and formed the community types prevalent today. The northern limit of these boreal forests began to retreat, reaching its present position about 4,500 BP, and the forest-tundra ecotone reached its maximum extent, similar to its modern position.

We can probably reconstruct the vegetation of the late Holocene of this area most effectively by examining the most recent pollen zone (T5-4 and SL-4) and then going back in time to the spruce woodland reconstruction we have just proposed for the mid-Holocene. The corrected diagrams (Figure 15) are helpful in this procedure. Both show that the modern dwarf shrub tundra landscape has prevailed unchanged since about 3,500–4,000 yr BP. Between 4,000 and 5,000 yr a transition occurred when the mid-Holocene continuous spruce woodlands diminished sharply and were replaced by alder-birch-spruce forest tundra, represented at T-5 by zone 3 to 4 and at Sleet Lake by the spectra between levels 200 and 100 cm. By 4,000 yr BP the spruce trees had disappeared and only scattered krummholz individuals persisted, as they have to the present day. Corroboration of the above reconstruction comes from pollen influx values of spruce in the SL sections (Figure 14). Spruce influx rose sharply from low values at about 10,000 yr BP to maximum values between 9,000 and 8,000 yr BP, and declined steadily between 8,000 and 6,000 BP. At 4,500 there was a sharp decrease to modern values. The decreasing spruce influx between 9,000 and 5,000 at Sleet Lake (Figure 14) could be interpreted as due to local opening or thinning of the otherwise continuous spruce canopy; we will explore this question further in Chapter 6 when we consider the alder pollen data.

Tuktoyaktuk 5

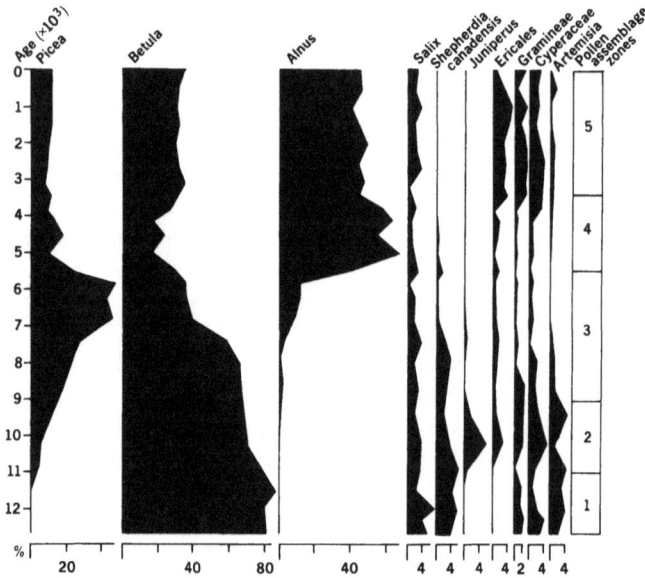

Sleet Lake

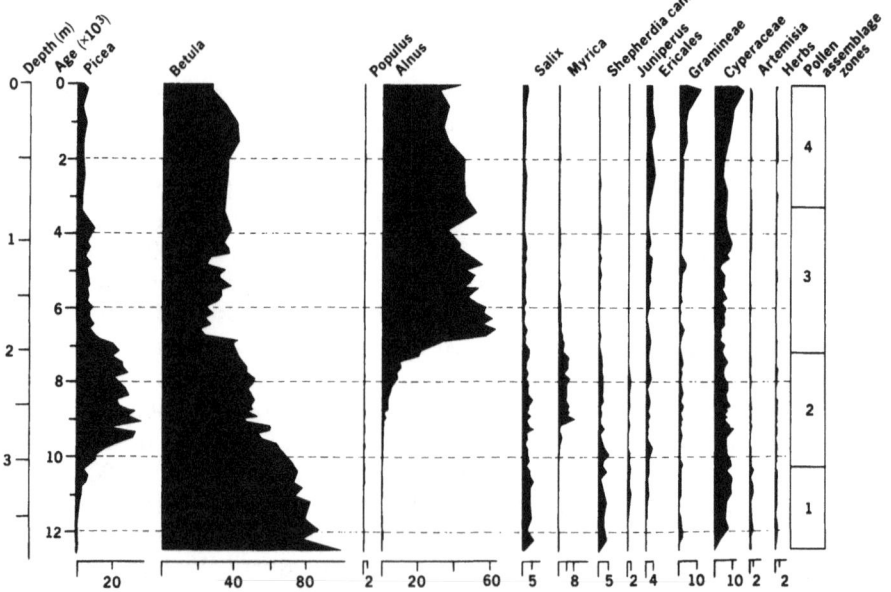

Figure 13 Summary percentage pollen diagrams of the Tuktoyaktuk-5 site (upper, after Ritchie and Hare 1971) and the Sleet Lake site (lower, after Spear 1983). A listing of all radiocarbon determinations used in these and subsequent diagrams can be found in Appendix 4.

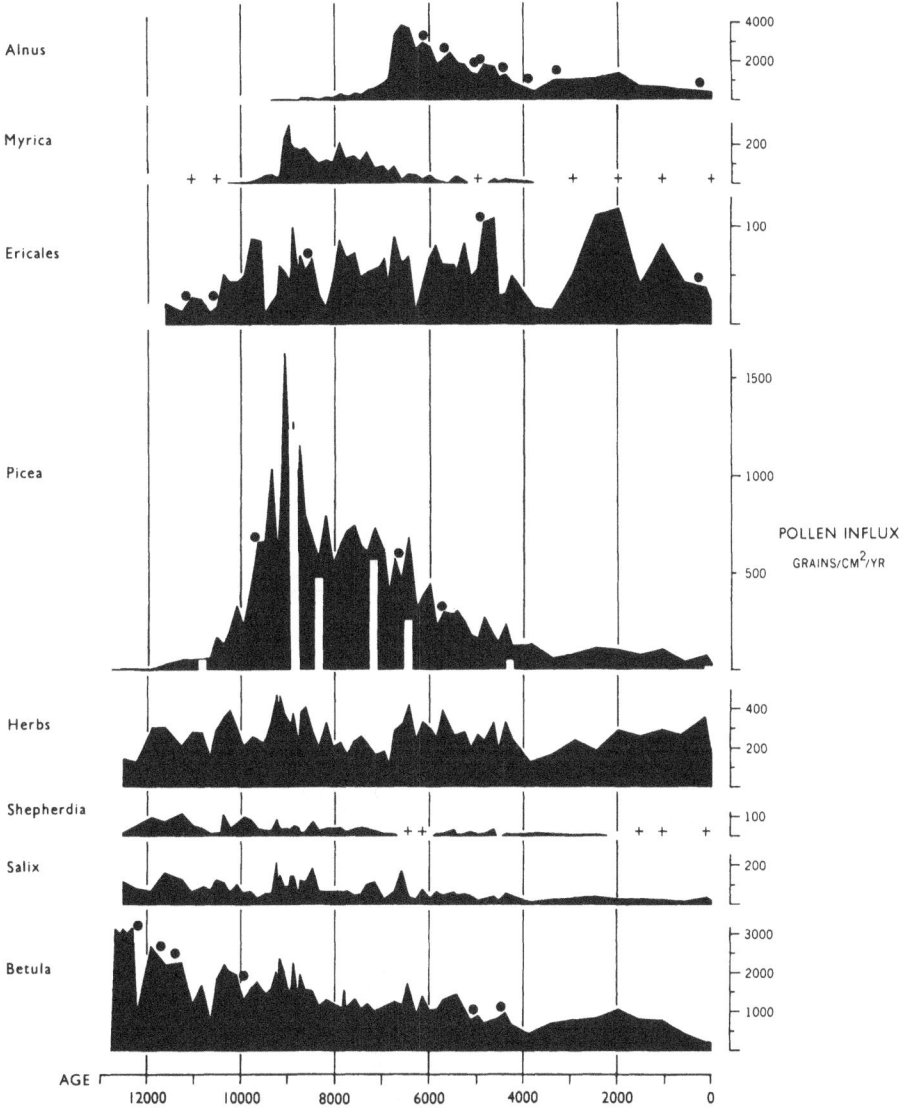

Figure 14 Influx values of the main pollen types, plotted against radiocarbon age, after Spear (1983). Note scale changes.

Tuktoyaktuk-5

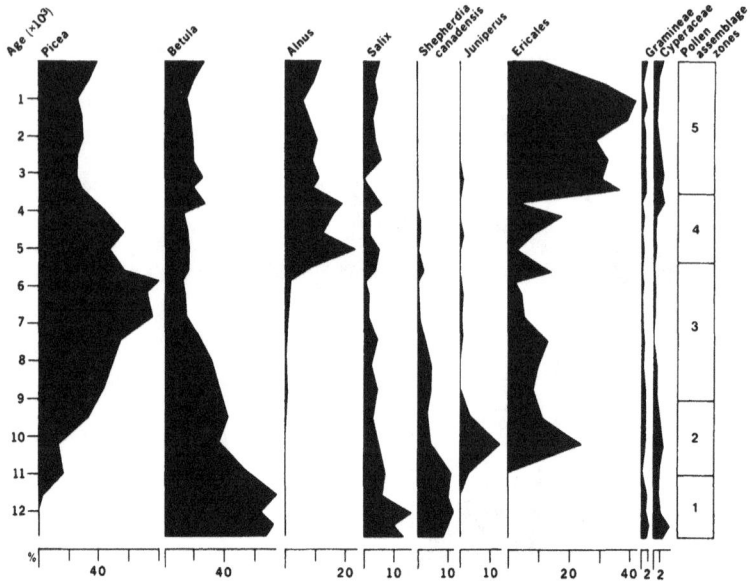

Sleet Lake

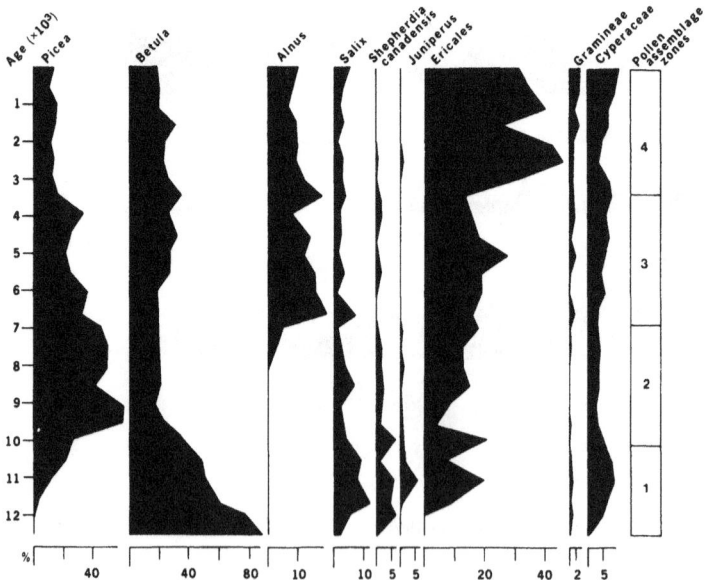

Figure 15 Summary percentage pollen diagrams for Sleet Lake and Tuktoyaktuk-5 based on corrected pollen totals, using the *R* values in Table 6

3.2. INUVIK AREA

A small area of outcropping bedrock 15 km southeast of the town of Inuvik includes several small, closed drainage lakes; we have cored three, of which two yielded informative results. These lakes are in karstic depressions in a complex of late Precambrian, Ordovician, and Devonian shales, limestones, and dolomites, comprising the Campbell uplift of the Aklavik Arch referred to in Chapter 2. The area was covered by an extension of Laurentide ice, and I am assuming here that it was a late-surging tongue of the Laurentide glacier that extended down (northward) the lower Mackenzie Valley about 14,000 BP, covered the Inuvik area, and terminated along the Tutsieta moraine of Hughes (personal communication) and to an uncertain extent northwards beyond Sitidgi Lake (Figure 6). Only a thin discontinuous veneer (~50 cm) of gravels, sands, and clay left by this late glaciation overlies the bedrock, but the surrounding terrain, consisting of fluvioglacial gravels and silts, shows pronounced fluted or obscurely drumlinized hills trending roughly north-south (Mackay 1963; Fyles et al. 1972).

Complete, similar pollen stratigraphic sequences have been produced from the two sites. One has been published in detail (Ritchie 1977); the other is being described here for the first time. The sites are referred to informally as Maria Lake and Twin Tamarack Lake, abbreviated hereafter as M-Lake and TT-Lake. M-Lake is 7.7 ha in surface area with a maximum depth of 22 m and is situated in the centre of the bedrock upland, while TT-Lake is 4 ha and 4.5 m in depth, at the northwest edge of the bedrock where it abuts on Pleistocene deposits. The vegetation of the upland has been described in detail (Ritchie 1977). Both sites are within the northern boreal woodland, roughly 15 km south of the limit of continuous spruce woodlands and 20 km south of the limit of trees. The chief cover of upland sites is vegetation of the type described in Chapter 4 as 'white spruce stands on uplands – calcareous parent material,' and the reader is referred to that section (Section 2.1(c)) or to the original, detailed publication (Ritchie 1977) for a description. TT site is bounded on its long east shore by an open spruce-larch cover of that type, but the west shore and hinterland are developed on a low, poorly drained surface consisting of a thin veneer of Pleistocene glaciofluvial silts and sands, covered by a *Picea mariana–Sphagnum* mire woodland, of the type described in Chapter 4, Section 2.1(i).

The pollen stratigraphy of the two sites is very similar (Figures 16, 17). The TT sequence has yielded a longer record than that of M-Lake, probably because the dispersant-and-sieving method of Cwynar et al. (1979) only became available after the M-Lake sediments were sampled and processed. The results from samples in the lowest silty clays with low pollen concentration (~20,000 grains / ml) enabled us to derive a useful record from sediment we were obliged to reject when the M-Lake material was analysed.

Four pollen assemblage zones and three subzones can be derived in both diagrams, by direct inspection of the percentage diagrams (Figures 16 and 17).

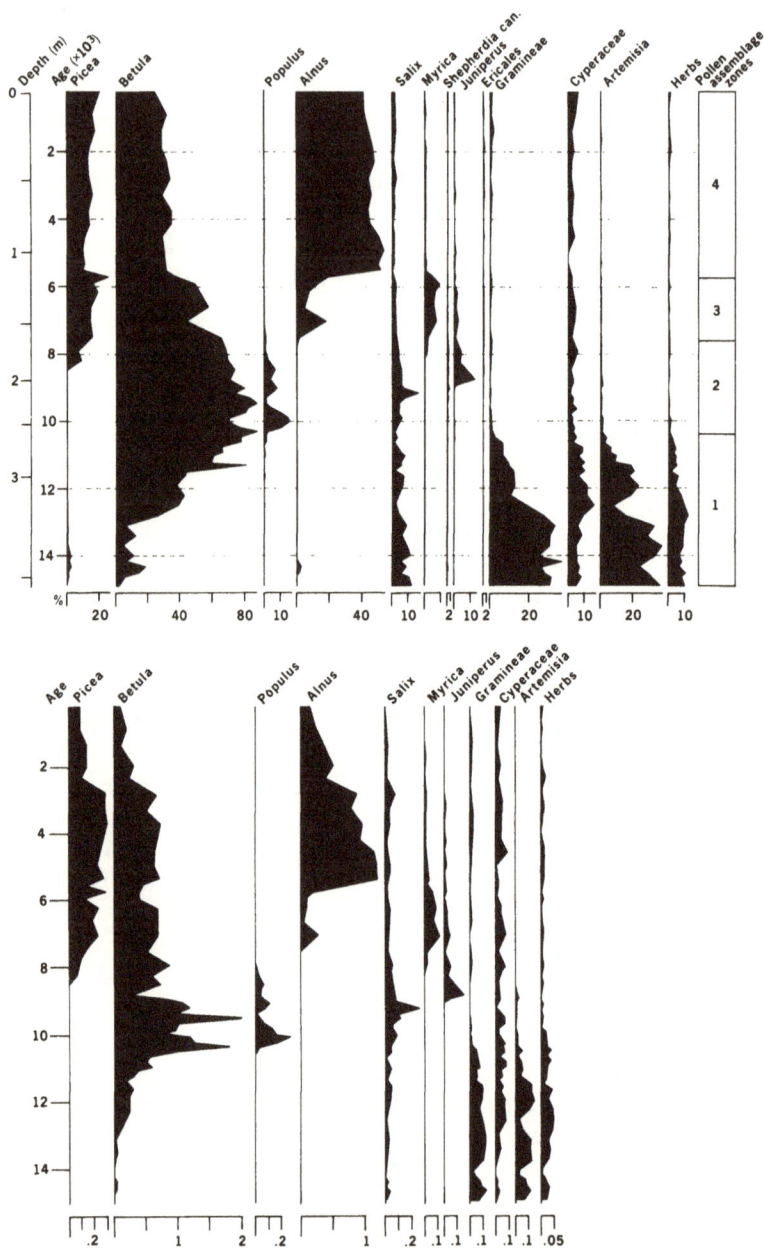

Figure 16 Summary percentage and influx pollen diagrams for the Twin Tamarack site. Note scale changes in influx, shown as grains per cm^2 x 10^3.

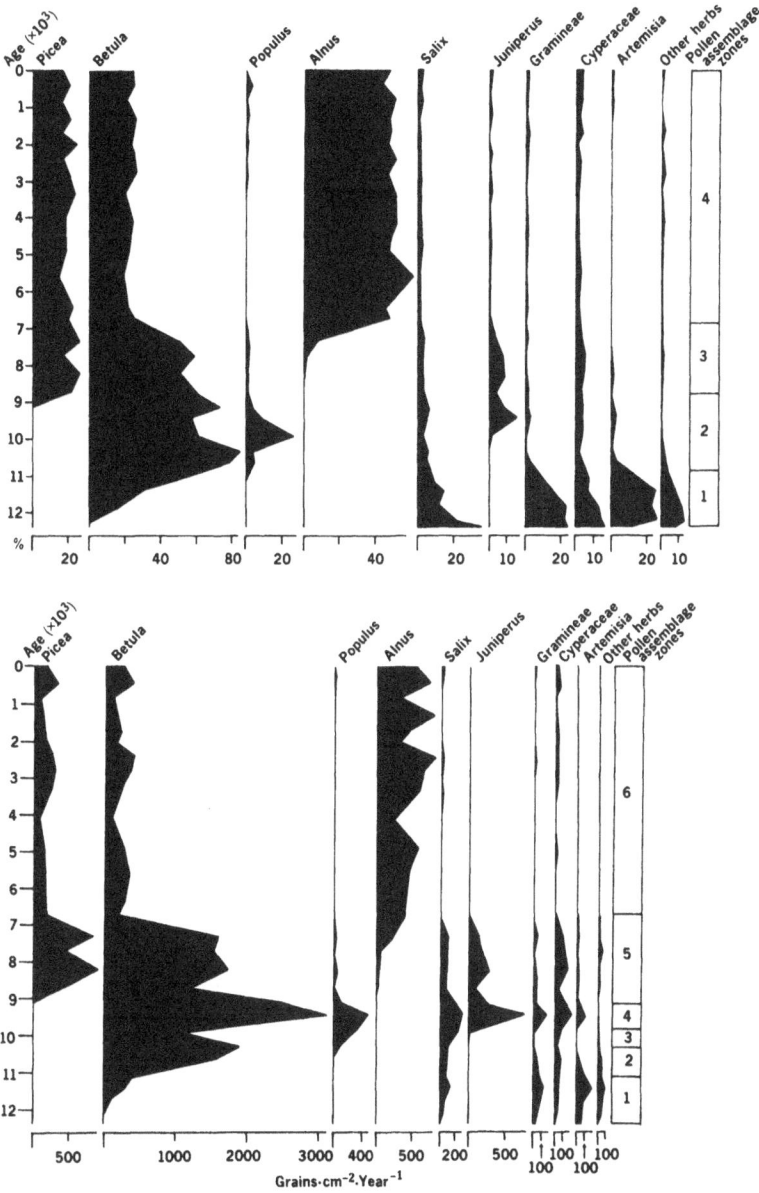

Figure 17 Summary percentage and influx pollen diagrams for the M-Lake site. Note that pollen zone 2 in the percentage diagram is subdivided into three subzones in the influx version, distinguishing the birch, birch + poplar, and birch + poplar + juniper assemblages.

Assemblage zone 1 is better represented at TT than at M, for the reasons stated above, and it is characterized by *Salix* (5–10%), *Artemisia* (~20%), and Gramineae (~15%) with a variety of herbs in low frequencies but of ecological interest: *Plantago canescens*, *Thalictrum*, *Oxytropis*, *Hedysarum*, *Saxifraga*, and *Polemonium*. It spans the 14-C interval from roughly 13,500 to 12,000 BP. Zone 2 is characterized by *Betula* (*glandulosa* type) pollen, and a rise in birch from low, erratic values to ~20% delimits the lower zone boundary. This main birch zone is subdivided into three subzones, identical at both sites, and collectively occurring between 12,000 and roughly 8,500 BP. Zone TT-2 ends at about 8,200 and M-2 ends at 9,000 BP. Subzone 2a is a *Betula*-herb assemblage, with 30–40% *Betula* (*glandulosa* type), ~5% *Salix*, and ~10% each of *Artemisia*, Gramineae, and Cyperaceae. Subzone 2b occurs at about 10,000 BP and is distinguished as a *Betula–Populus* assemblage with 70–80% birch and 15–20% poplar. The herb frequencies have become reduced to less than 3% for all taxa. Subzone 2c is characterized by a *Betula–Salix–Juniperus* assemblage, with birch frequencies still high (~80%) and the willow and juniper percentages rarely exceeding 10%.

Assemblage zone 3 is bounded below, at roughly 8,500 BP, by the increase of *Picea* from none or very low percentages to about 20%, and it is terminated at the level, ~6,000 BP, when birch declines and alder rises steeply to maximal frequencies. The zone has, in addition to *Picea* at ~20%, *Betula* (50% shrub type) frequencies from 40 to 50% and low values of *Salix* and *Juniperus*. The percentage diagrams show little change in the pollen frequencies throughout zone 4, with *Picea* (~20%), *Betula* (~25%), and *Alnus* (~50%) the dominant types.

The influx diagrams (Figures 16 and 17) indicate that the zonation based on percentages is reasonably accurate. They also show clearly that the percentage diagrams give an inflated impression of the non-arboreal pollen frequencies in the late-glacial, and the longer TT site record in particular illustrates that total influx values of 350 cm^{-2} yr^{-1} are rarely exceeded before dwarf birch occurs, at ~12,000. The TT diagrams also provide evidence that the dwarf birch percentage curve gives a spuriously early age (13,000) for the major increase, which, according to the influx curve, actually occurred at 11,000.

Vegetation Reconstruction
When we turn to reconstructing the past vegetation of these sites from the pollen record, some formidable difficulties present themselves. The first is that modern pollen spectra from lake sediment in different zonal vegetation types in this area, as simple as tundra, forest-tundra, and forest, cannot be matched with some fossil pollen assemblages, either by direct inspection or by numerical analyses (Ritchie 1974, 1977). Part of the reason for the apparent absence of numerical analogues of certain fossil assemblages is that a few abundant types (*Alnus* and *Betula*) exercise undue effect on the correlations and resulting vectors, but so far no satisfactory method of weighting samples has emerged.

However, the calculation of a relative diagram based on pollen sums corrected for over- and under-representation of the main pollen types (*R*-values), as we did for the Tuktoyaktuk sites, yields a sequence that suggests the vegetation changes more directly than the original, and this diagram (for the TT site) together with the concentration and influx curves for the important taxa forms the basis of the following vegetation reconstruction.

R-values derived from quantitative vegetation analyses reported in Ritchie (1977, Table 6) and modern pollen data were used to produce corrected percentages (Figure 18), and that diagram together with the pollen influx diagrams (Figures 16 and 17) forms the basis of this attempt to reconstruct the past vegetation of this rock upland. The corrected TT diagram reduces the sharp zone boundary (1–2) at about 12,000 BP by reducing the birch frequencies and increasing those of the herbs. It is clear that the first vegetation to occupy the area after the Mackenzie Valley tongue of Laurentide ice disintegrated at about 13,500 BP was an open, herb tundra. The low initial influx values suggest a sparse community, but the rising influx values during this episode and the increase in birch frequencies and influx indicate a slowly increasing tundra cover.

Between 11,000 and 10,000 BP dwarf birch expanded rapidly at the expense of some of the herb component until it occupied roughly 20% of the ground cover. During the same interval, poplar and willow expanded sharply, and the former was the first tree species to occupy a significant place in the vegetation of the area. Both sites register this sequence of events, with close agreement between the radiocarbon dates. Shortly after, at about 9,500 yr BP, the ground vegetation changed in composition with further decreases in total herb cover in response to the spread of juniper and heaths. Roughly one millennium later spruce arrived and expanded to about the level of its modern representation in the vegetation. It was accompanied somewhat later (about 7,000 BP) by larch and tree birch and finally at 6,500 BP alder arrived and assumed its role in the upland (*Alnus crispa*) and lowland vegetation (*A. incana*).

It is of interest to compare the data from the M-Lake and TT sites with the two continuous complete records from the Tuktoyaktuk area. Some differences can be ascribed to latitude: for example, the most recent pollen zones reflect clearly the differences between sites in the tundra from those in the woodlands. The tundra sites show a greater change in mid-Holocene time, reflecting the marked reversal from a spruce woodland to a treeless landscape, while at the Campbell Hills sites it is clear that less pronounced changes in vegetation occurred. Most interesting are the comparisons between the site pairs during the few millennia immediately following deglaciation. The chief difference between these contemporaneous tundra landscapes, separated by only 80 km, was that the Tuktoyaktuk Peninsula uplands were covered in a birch shrub-tundra with abundant willow and soap-berry but quite localized herb communities while the Campbell Hills supported an open herb tundra with some willows but very little dwarf birch

M Lake

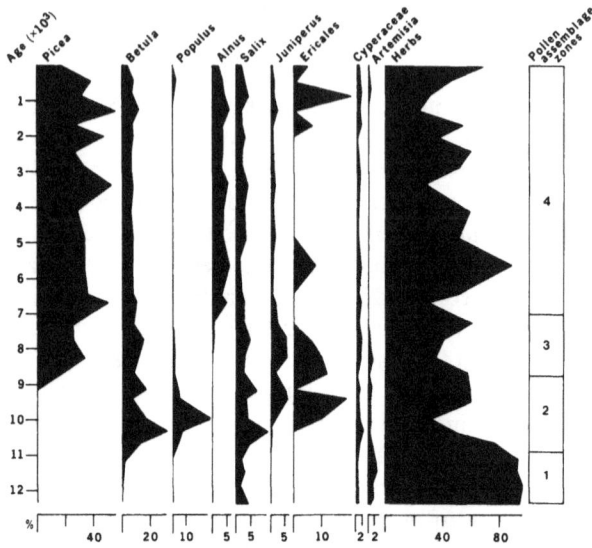

Twin Tamarack

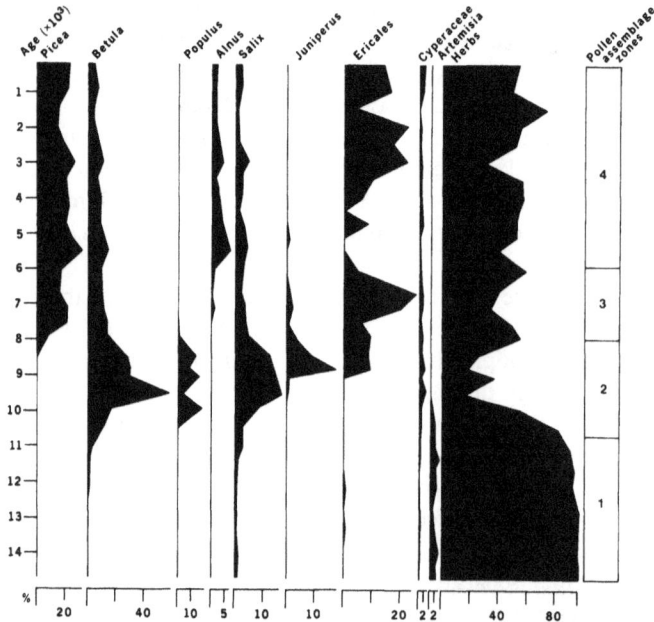

Figure 18 Summary percentage diagrams for Twin Tamarack and M-Lake, based on corrected pollen totals after the *R* values in Table 6 were applied to the original sums

and no evidence of soap-berry. It is possible that these differences were due to the chemistry of the substrata, and particularly to the presence of abundant limestone in the Campbell Hills area, which tends to inhibit dwarf birch and permit an expansion of herbs. An alternative explanation is that the early records from the Tuktoyaktuk Peninsula, showing a dominance of dwarf birch, are an unreliable registration of the initial vegetation because the sediments deposited during the earliest interval were of thermokarst origin. We have already noted the interpretation by Fyles et al. (1972) that the lakes probably originated as thermokarst collapse basins during the initial late-glacial warming. However, the absence of sediment discontinuities and the regular sequence of radiocarbon ages would not be expected from sediments derived primarily from thermokarst collapse. A tabular summary of the pollen stratigraphy and suggested vegetation (Table 7) will facilitate direct comparison of the four sets of data.

The final site to be considered from the Inuvik area is the Twin Lake peat profile near the town of Inuvik, on the bank of the east channel of the Mackenzie River, at 69°12′ N, 132°27′ W. This 3 m exposure of postglacial peat was described and sampled by Mackay (1963), and the results of pollen and radiocarbon analysis were reported by Mackay and Terasmae (1963). Their results indicated a radiocarbon age of the peat of 7,400 yr (GSC-16) and the pollen diagram indicates a lowermost spruce-birch zone with a high proportion of non-arboreal pollen, followed by a spruce-birch-alder zone near the top of the section. The site was resampled by M. Kuc, reported in Fyles et al. (1972) in the form of a large table showing depth, sub-fossil occurrences, radiocarbon ages, and attempts to correlate with Mackay and Terasmae's (1963) results, as well as those from Ritchie and Hare's (1971) T-5 site. The result is confusing. The table indicates, apparently on the basis of spruce macrofossils, that spruce forest invaded the area 11,500 years ago. Records of alder cones and leaves are shown from sediment dated at older than 8,200 yr, although it is not made clear whether all samples contained both types of macrofossil or not.

Matthews (1975b) and Hopkins et al. (1981) accept these results, although they note that they are an example of 'the incongruence of macrofossil and pollen evidence' because of the apparent absence of alder pollen from sediments yielding alder cones. The version of the biostratigraphy at Twin Lakes published in Fyles et al. (1972) is at variance with the pollen and spruce macrofossil results from three lake sites near Inuvik (M-Lake and TT-Lake reported here, and a new site to be reported on in detail in a later publication and known informally as Kate's Pond) and with all other results from sites in the Mackenzie Delta region, as reported in this chapter. It appears to me more likely that the Twin Lakes data reported in Fyles et al. (1972) are incorrect and that the Mackay and Terasmae (1963) version is accurate, primarily because the growing volume of biostratigraphic data from nearby sites correlates closely with the latter version. We will return to the question of apparent anomalies in pollen and macrofossil data when we examine the alder record, in Chapter 6.

Table 7 A comparative summary of the pollen stratigraphy and vegetation reconstruction for the Tuktoyaktuk and Campbell Hills, Inuvik area

Age (14-C yr)	POLLEN STRATIGRAPHY		VEGETATION		POLLEN STRATIGRAPHY		Age (14-C yr)
	T5	SL	Tuktoyaktuk Peninsula	Campbell Uplands	M-Lake	TT	
1—	*Picea* ~ 12% *Betula* 25–30% *Alnus* 40% *Ericales* 2–4% Herbs 5%	*Picea* ~ 12% *Betula* 20–30% *Alnus* 40% *Ericales* 8% Herbs 4%	Dwarf birch–ericad–sedge cotton–grass tundra as today: spruce absent or krummholz: outliers of balsam poplar	Open spruce woodland with larch, tree birch, and dwarf shrub, herb, lichen ground cover as at present	*Picea* 15–20% *Betula* 20–25% *Alnus* 40–50% *Juniperus* 5% Herbs 5%	*Picea* 15–20% *Betula* 20–25% *Alnus* 40% *Juniperus* 5% Herbs 4%	—1
2—							—2
3—							—3
4—	Zone 3: *Alnus* ~ 60% *Betula* ~ 20%	Zone 3: *Alnus* ~ 55% *Betula* ~ 20%	Transitional forest tundra with increase of alder, spread of dwarf shrub tundra	As at present with more dense cover of spruce, with tree birch, larch, and poplar. Alder local			—4
5—	*Alnus* rise	*Alnus* rise					—5
6—	Zone 2: *Picea* decline	Zone 2: *Picea* decline	Continuous coniferous woodland on all upland sites, dominated by spruce with local occurrences of poplar, larch, and, later, tree birch		*Alnus* rise	*Alnus* rise	—6
7—	*Picea* peak (40%) *Myrica* < 2%	*Picea* peak (30%) *Myrica* 5–6%		Spruce woodland with dwarf shrub–herb ground cover: occasional larch, poplar	*Picea* 20% *Betula* 40% *Juniperus* 4% *Picea* rise	*Picea* 20% *Betula* 50–60% *Picea* rise	—7
8—							—8
9—	*Picea* rise	*Picea* rise		Dwarf birch–herb tundra with juniper, ericads, and poplar	*Betula* ~ 50%; *Populus* 10% *Juniperus* 10%; *Salix* 10%	*Betula* > 60%; *Populus* ~ 8% *Juniperus* 5%; *Salix* 5%	—9
10—	Zone 1: *Betula* (dwarf) 80%	Zone 1: *Betula* (dwarf) 90%	Dwarf birch–dominated tundra with frequent local *Shepherdia*, and *Salix* –sedge carr or marsh vegetation in lowlands. Localized herb tundra on xeric sites		*Betula* ~ 60%; *Populus* 20%	*Betula* 70%; *Populus* 15%	—10
11—	*Salix* and *Shepherdia* ~ 4%	*Salix* 5–7% *Shepherdia* 4% Herbs 12%		Herb tundra with birch	*Betula* 50%, *Salix* 5%, herbs 30%	*Betula* 50%; *Salix* 5%, herbs ~ 25%	—11
12—							—12
13—	Herb total 10%			Open herb tundra	Herb total ~ 70% *Salix* 20%	Herb total ~ 70% *Salix* 10–20%	—13

3.3. TRAVAILLANT LAKE AREA – SW-SITE

Recent results, not yet published, from a small (8 ha), shallow (5 m) pond referred to informally as Sweet Little Lake (hereafter referred to as SW-site) provide valuable evidence from a poorly known part of this region. The pond is in rolling morainic terrain immediately to the west of Travaillant Lake, and lies at 67°39′N, 132°01′W, approximately 10 km west of the Tutsieta moraine.

The modern vegetation consists of continuous boreal forest on all upland sites, dominated by *Picea glauca* and *Betula papyrifera*, with local dominance of *Picea mariana* associated with *Larix laricina* on peaty soils, *Populus tremuloides* on well-drained slopes, often showing evidence of recent fires, and *P. balsamifera* localized on heavy till soils and along streams.

A 4 m core was recovered of which the upper 376 cm produced satisfactory pollen results; six 14-C age determinations were made from samples of the core. The pollen stratigraphy (Figure 19) shows three main assemblage zones. The lowermost, spanning the radiocarbon interval from 10,500 to 9,000 yr BP, is dominated by dwarf birch (60–70%), poplar (5–10%), *Shepherdia canadensis* (from over 20% in the lowest samples to 2%), and *Juniperus* (2–10%). The second, from 9,000 to roughly 5,500 yr BP, is a zone of dwarf birch dominance (50–60%) associated with *Picea* (10–20%); *Populus* declines to negligible amounts at about 8,500 and *Juniperus* slightly later. The modern spectrum is established at about 5,000 yr BP, represented by a *Picea* (16–22%), *Betula* (20–25%), and *Alnus* (40–50%) zone.

The influx diagram confirms the general zonation scheme for this site and in addition provides a more accurate record of the pollen input of individual taxa, particularly *Betula*, *Alnus*, and Gramineae. Preliminary pollen size statistical analyses using the method of Birks and Peglar (1980), adapted with a data set of modern samples from northwest Canada by Spear (1983), indicate that *Picea glauca* was the dominant spruce at this site from 9,000 to 8,000 and that *P. mariana* assumed that role at about 6,500 and has retained it to the present day. The ratios of white spruce to black spruce pollen in SW-Lake samples are as follows, the radiocarbon age being shown in brackets: 4:1 (9,000), 1:1 (8,000), 1:4 (6,500), 1:4 (5,000), and 1:9 (4,500).

There is broad agreement between these results and the diagrams from M-Lake and TT-Lake 90 km to the north-northwest near Inuvik. The chief differences are the younger age of the SW section and the absence there of a late glacial herb-dominated zone. The suggested vegetation reconstruction is as follows. At about 11,000 yr BP, following the disintegration southward of the lobe of Laurentide ice that had extended northward slightly beyond the distal end of the modern Mackenzie Delta, shrub tundra vegetation colonized the area from the immediate north and possibly west where it prevailed on most upland, non-calcareous surfaces. The chief community on mesic upland sites was a *Betula glandulosa–Salix* shrub tundra. Local shrub facies dominated by *Shepherdia*

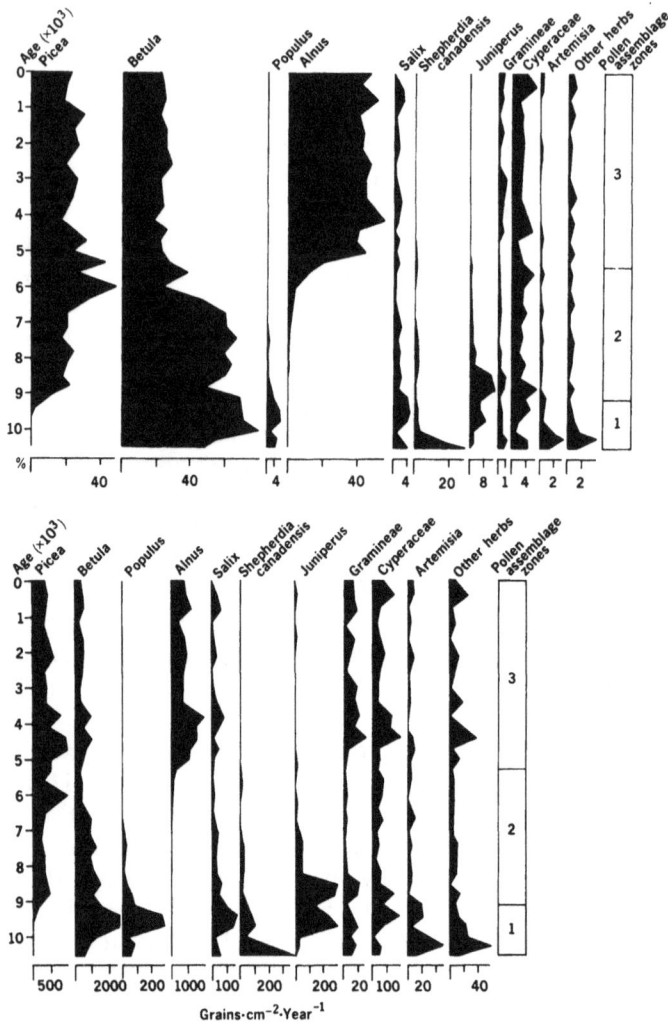

Figure 19 Summary percentage and influx pollen diagrams for Sweet Little Lake. Note the scale changes.

canadensis occupied more xeric habitats, on steep slopes and knoll summits, while clonal groves of *Populus balsamifera* were frequent at lakeshores and along streams. *P. tremuloides* might have been present, forming stands on well-drained upland soils. *Juniperus communis* expanded into these communities occupying similar open, well-drained habitats to those of *Shepherdia canadensis*. The arrival of *Picea glauca* at roughly 9,000 BP was followed by a rapid expansion of spruce forest onto upland sites, progressively repressing the *Populus*, *Juniperus*, and *Shepherdia* as open, unshaded habitats diminished. The question of the source of this spruce migration will be discussed later in Chapter 6 in the general context of migration. *Picea mariana* arrived with or shortly after *P. glauca* but did not increase in abundance until 8,000. Later *Alnus* (*crispa* and *incana*) arrived and expanded between 6,000 and 5,000 yr BP; the modern mosaic of boreal forest communities was essentially in place by 5,000 years ago. Competition from white spruce and the paludification of lowland sites and stabilization of lakeshore habitats probably caused the diminution of *Populus balsamifera* at about 8,500 yr BP, but that does not explain the near absence of poplar pollen from modern and late Holocene levels in the sediment although the tree occurs frequently in the area today.

3.4. NORTH YUKON

(a) *Hanging Lake*
The longest continuous pollen record from primary sediment is from Hanging Lake, analysed and reported by Cwynar (1982): it provides a reconstruction for roughly the past 30,000 years. The following is based entirely on the work of Cwynar (1980, 1982). The lake is on a ridge of shale and sandstone 10 km southwest of the Barn Mountains, at 68°23′ N, 138°23′ W (Figure 12). It is a closed-drainage basin roughly 60 ha in area and with a maximum water depth of 9.5 m, and a small outlet stream drains from the northeast end. It lies beyond treeline in an area of tundra. Cwynar (1980) provides a detailed description of the four main community types that prevail – *Eriophorum–Betula* tussock tundra, sedge meadows, heath meadows, heath tundra – and a fell-field discontinuous dwarf shrub tundra.

A 403 cm core of sediment was sampled in duplicate, and 21 radiocarbon age analyses provide a secure chronology for pollen influx estimates. The oldest pollen zone (HL-1) is the Herb Zone spanning 33,000 to 18,450, and it is dominated in the percentage diagram (Figure 20) by Gramineae, *Artemisia*, and Cyperaceae. However the influx data (Figure 21) show that herb pollen totals are very low (5–50 grains cm^{-2} yr^{-1}), and several of the taxa that are prominent in the percentage diagram in fact show their highest influx values much later, between 11,000 and 8,000 yr. He separates a *Picea* subzone at 21,600–18,450 on the basis of a percentage increase in spruce, birch, and alder. Zone HL-2, the *Salix*-Cyperaceae zone from 18,450 to 14,600 yr, is distinguished by increases in both the relative frequencies and the influx values of willow and sedge pollen types, but again maximum influxes for these taxa in the entire section occur later, in the early Holocene. Zone HL-3 is a *Betula* zone, from 14,600 to 11,100 yr, and there is a ×5 increase in birch influx paralleling the rise to values of 60–80%, but a much larger influx is recorded at about 11,500 yr. An ericales zone (HL-4) occurs at 11,100–8,900 indicated by marked increases in both influx and percentage of ericad pollen, while increases of *Picea, Populus, Alnus*, and *Equisetum* occur in this zone, and *Betula* values (influx) reach their maximum values. Zone HL-5, an

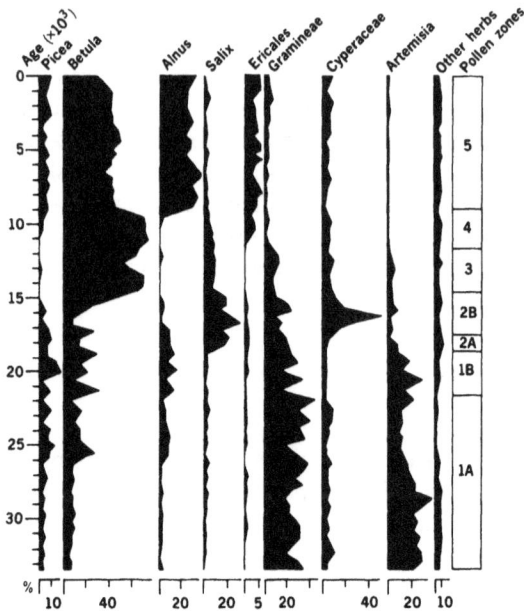

Figure 20 A simplified percentage pollen diagram of the Hanging Lake site, showing the main pollen types plotted against radiocarbon age (after Cwynar 1982)

Alnus crispa zone, begins at 8,900 yr with a rise of alder frequencies to about 35%, and no significant changes occur up to the present day except that the influx values of all taxa show a gradual diminution.

Cwynar interprets the herb pollen zone evidence as follows (1982, pp 13–14):

The low total pollen influx, ranging from 5–100 grains cm^{-2} $year^{-1}$, the low organic content of the sediment, the prominence of pre-Quaternary spores presumably derived from weathering of the shale bedrock basin, and the occurrence of open ground taxa such as *Dryas, Saussurea, Phlox, Polemonium, Bupleurum triradiatum, Lesquerella arctica, Androsace, Saxifraga tricuspidata*-type, and *Thalictrum*, all indicate that the vegetation was sparse. The small number of herb pollen types [*S* varies from 15 to 10] suggests it was also depauperate in comparison with the modern vegetation.

The influxes of Gramineae and *Artemisia* – whose relative abundance has been used to infer a grassland or steppe-tundra vegetation by many authors – are no more abundant in this zone than during subsequent ones. The only *Artemisia* at Hanging Lake today is *A. arctica* ssp. *arctica* which is common along the eastern fell-field ridge (Figure 3e). This is the same species that is responsible for the high *Artemisia* percentage in modern pollen spectra (Table 3) from the Old Crow Range. It may be the main source of *Artemisia* pollen throughout the Hanging Lake section but some of the *Artemisia* pollen is likely derived from other species found in rocky uplands of the northern Yukon today, such as *A. glomerata, A. globularia*, and *A. furcata*. The grass pollen is probably derived both from

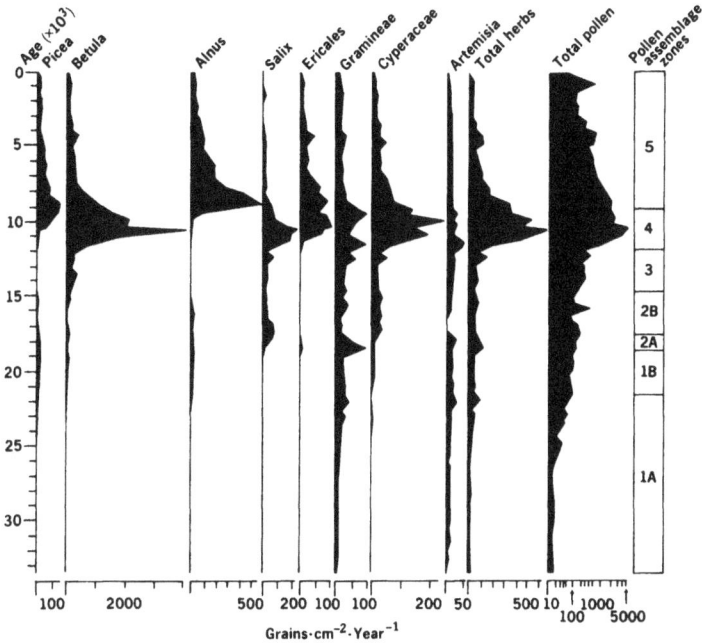

Figure 21 Influx values of the main pollen types at the Hanging Lake site, plotted against radiocarbon age (after Cwynar 1982)

upland taxa, such as *Poa glauca*, *Hierochloe alpina*, and *Festuca brachyphylla*, that are common today in the discontinuous tundra communities of the drainage basin, and emergent aquatic grasses – principally *Arctophila fulva* which grows occasionally in the lake today but does not form any large marshy stands because the bedrock basin drops off relatively quickly and the shoreline is scoured by rafting ice during the spring.

He offers the tentative suggestion that the *Picea* subzone, coinciding with the beginning of organic sedimentation, represents an interstadial, possibly correlative with other evidence for buried soils dated at about 21,000 yr BP from the Brooks Range of Alaska (Hamilton 1979). He goes on to suggest that spruce might have persisted in the northwest in highly localized sites, throughout the latest glacial cycle. We will return to this question and to the full-glacial evidence presented by Cwynar (1982) in the following chapter.

Zone HL-2 is interpreted as a plant cover little changed from zone HL-1 except for an expansion of snow-bed willow and sedge communities.

Zone HL-3 is interpreted as a time of steady increase in species richness, the development of a more continuous cover of tundra with local expansion of dwarf birch into favourable habitats. Zone HL-4 is described as a period of rapid development of 'wet-mesic heath communities,' with a marked expansion of dwarf birch and willow to form low shrub heath tundra similar to modern communities in the area. Cwynar notes that the beginning of this zone (~11,000) also has maximum influx values of poplar, *Typha latifolia*, and *Myrica gale*. Finally, zone HL-5 is interpreted as reflecting little change in the vegetation except for an expansion of alder to its modern status.

In summary, these results indicate that the modern pattern of plant communities at Hanging Lake became established about 8,000 BP. The heath elements in the modern vegetation expanded during the two preceding millennia. Prior to 11,100 BP, during the closing stages and the maximum of the latest glacial cycle (from about 25,000 to 11,100 yr), the vegetation was a mosaic of tundra communities similar in structure and composition to modern arctic vegetation north of the low-arctic shrub-heath zone.

(b) *Lateral Pond*

The South Richardson Mountains area was chosen in the search for polleniferous sediments on the basis of a suggestion by Hughes (1976, personal communication) that moraine-dammed lakes existed there and that the moraines probably were formed by the ultimate or penultimate advances of Laurentide ice. Six lakes in the Doll Creek area were examined and sounded, three were cored, but only one produced a continuous record of primary sediment that spans the glacial-interglacial transition. It is Lateral Pond, informally so named because it lies between morainic ridges along the north side of the Doll Creek valley, which empties into the Peel River drainage. Lateral Pond is small (3.25 ha), shallow (3.4 m), and lies at 620 m above sea level close to the western flanks of the southern extremity of the South Richardson Mountains (Figure 12), at 65°27′N, 135°56′W.

The account that follows is a summary of a detailed description of the modern vegetation and pollen stratigraphy of the site (Ritchie 1982). The mountainous terrain and mixture of calcareous and non-calcareous sedimentary bedrock types support a rich variety of tundra and forest plant communities. Pollen percentage and influx data (Figures 22 and 23), supplemented by numerical zoning techniques (Figure 12 in Ritchie 1982), show clearly that the pollen diagram is divisible into a lower section, between ~16,000 and 12,000 yr BP, and an upper from 12,000 BP to the present. The former consists of non-arboreal taxa occurring with a low total influx (50–500 cm^{-2} yr^{-1}) and the latter of largely arboreal taxa with total sample influxes of 1,000–2,500 cm^{-2} yr^{-1}. The alder unit has 30 herb taxa, occurring in low concentration in all samples, made up predominantly of species or genera whose modern occurrences are in arctic-montane tundra communities (e.g. *Aconitum, Astragalus, Dryas, Hedysarum Oxytropis, Phlox, Polemonium, Polygonum viviparum*, and species of *Saxifraga*). There is an increase in *Salix* influx at about 13,000 yr BP, and then quite rapid fluctuations in the influx of herbs, grasses, and willows occur at about 12,500 yr. The vegetation reconstruction derived is of a 'sparse cover, probably a mosaic of chionophilous communities, local rich swards on south-facing middle and lower slopes where both warmth and moisture were relatively abundant, and grass-sedge marsh communi-

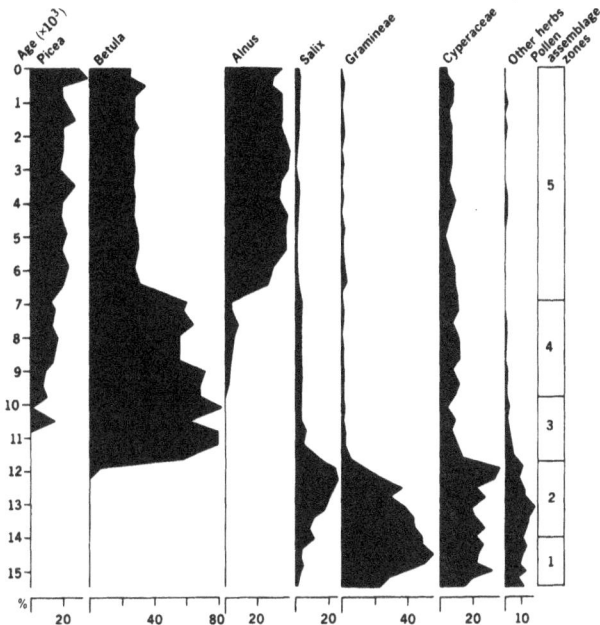

Figure 22 Summary percentage pollen diagram for Lateral Pond, based on Ritchie (1982)

ties with local willows in the Doll Creek bottomland soils with high water table and fine texture. Differences in bedrock type would have caused community composition contrasts similar to those of modern tundra vegetation' (Ritchie 1982). The use of an area-equivalent technique to estimate the approximate relative representation of elements or taxa in the vegetation (Ritchie 1982 and Figure 24 here) helps the reconstruction process. This approach calibrates the influx values by correcting for over- and under-representation by certain pollen taxa, and although there are several errors in the method (discussed in general in Birks and Birks 1980, pp 219–27, and for this site in Ritchie 1982, pp 569–71), the reconstruction derived in this way is not significantly different from what I proposed earlier by direct inspection of the influx and percentage diagrams (Ritchie and Cwynar 1982). They indicate that the sparse herb tundra on mesic sites expanded and became more dense during the period 16,000–13,500 yr BP, following which a decline of herbs and grass and an increase of willow cover reflected an expansion of willow scrub into hydric, bottomland habitats at the expense of marsh grasses. The transition to the upper arboreal-pollen zone at 12,500 is marked by large changes in influx values, suggesting instability of populations.

The upper unit, from 12,000 to 0 yr BP, represents a marked change in the vegetation from a herb-tundra landscape to one closely resembling the modern

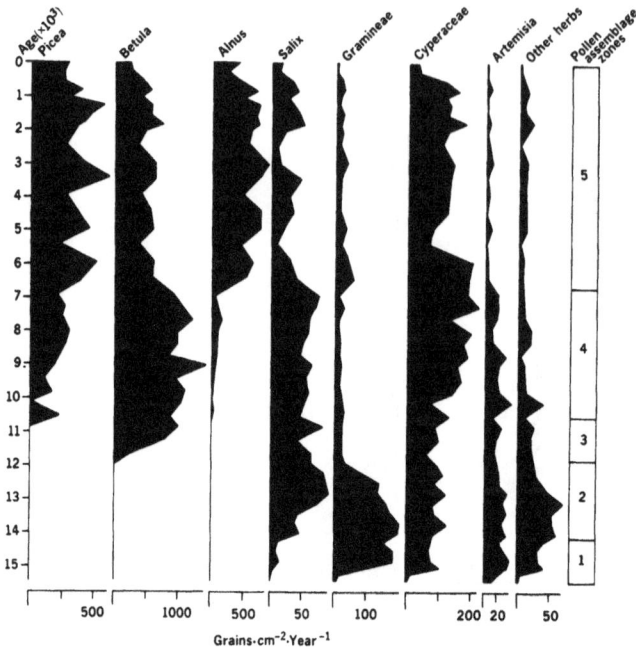

Figure 23 Summary pollen influx diagram for Lateral Pond, based on Ritchie (1982)

situation. That is, dwarf shrub tundras, particularly on non-calcareous substrata, expanded rapidly, dominated by *Betula glandulosa*, *Arctostaphylos*, *Vaccinium* ssp., *Cassiope*, and other heaths, while spruce woodlands spread onto surfaces at lower elevation, reaching approximately their modern extent by about 7,500 BP. Minor adjustments to this pattern occurred during the Holocene when tree birch was registered at about 6,000 yr BP but it never played a more significant role than it does today; it is confined to the non-calcareous sand-gravel soils of moraines in valley bottoms. Similarly, *Alnus* became established about 7,000 yr BP.

The Hanging Lake and Lateral Pond pollen records are the most thoroughly investigated sites with primary lake sediment from the entire northwest American region, as measured by the numbers of taxa identified, the sampling interval, the pollen sums, the radiocarbon control over age and sedimentation rate, and the ancillary information on modern vegetation. It is interesting therefore to compare the results to determine the extent to which they correlate with each other, bearing in mind that Hanging Lake is in a low arctic area of rolling topography with slight relief, near the arctic coast, while Lateral Pond is in a montane region within the subarctic or northern boreal forest climate. The striking similarity between the results is the abrupt transition at about 11,500 yr BP from an older herb tundra with low influx values to a shrub-birch tundra (with spruce at Lateral Pond) with relatively high influx values. Both sites indicate that the full and late

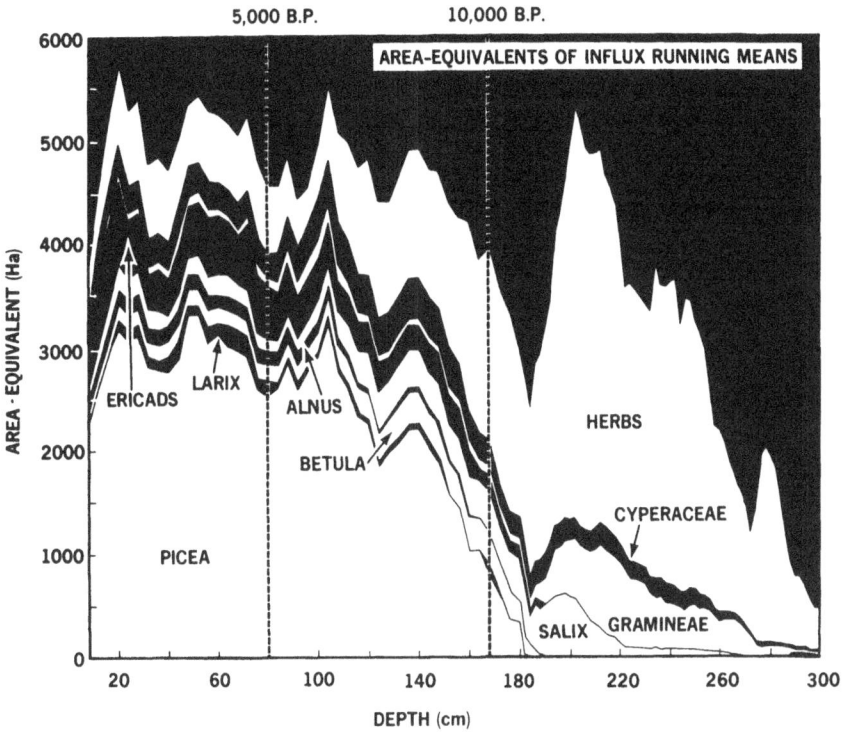

Figure 24 Five-sample running mean values of pollen influx for the main pollen taxa recorded at Lateral Pond, corrected by a factor to estimate the equivalent surface area occupied by each vegetation element, plotted against depth in the sediment and radiocarbon age (after Ritchie 1982)

glacial vegetation on uplands was a herb tundra that slowly, between about 14,500 and 11,500 yr BP, increased in cover and complexity. The Lateral Pond record begins after the recession of a glaciation whose ice cover was at least 100 km east of Hanging Lake.

(c) *Old Crow Flats*

Ovenden (1981, 1982) has completed detailed pollen, plant macrofossil, and peat stratigraphic investigations of a section of mire deposit in the Old Crow Flats. The site, informally designated Polybog, is situated near the southern periphery of the lowlands, 24 km north of the village of Old Crow, at 67°48′N, 138°47′W (Figure 12). The following account is based on the reports of Ovenden.

The mire is a large complex (1.0 by 0.3 km) of ice-wedge polygons developed in a deposit of roughly 2 m of peat and organic silts overlying the lacustrine clays of Glacial Lake Kutchin. These sediments are perennially frozen, and the active layer in late summer is 25 cm deep. A core 7.5 cm in diameter and 2.10 cm long

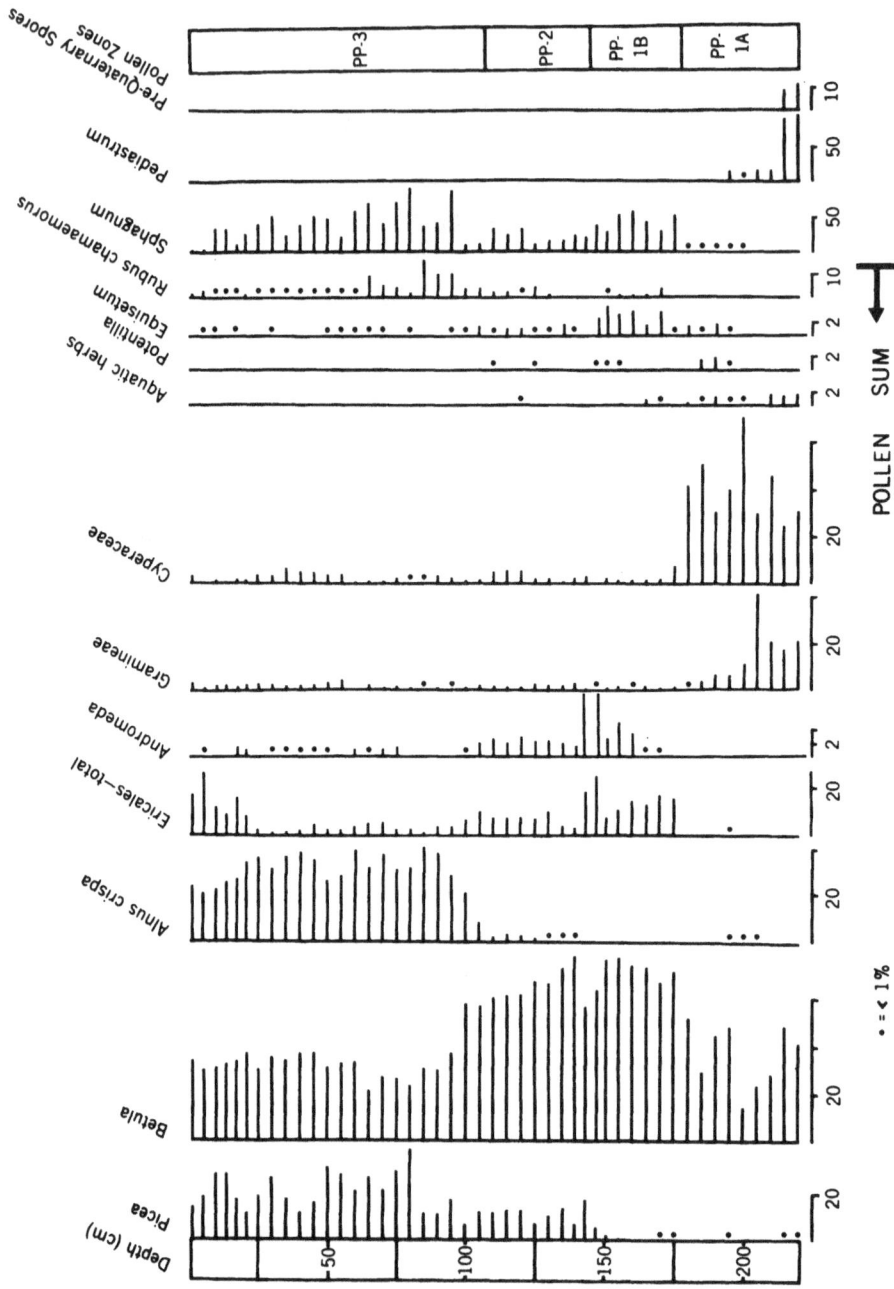

Figure 25 Percentage pollen diagram from Polybog site, after Ovenden (1982)

was raised in segments by a modified Sipre corer (Hughes and Terasmae 1963). Eleven samples were analysed for radiocarbon age, and pollen, plant macrofossil, and peat analyses were completed on the rest of the core.

The pollen analysis yielded a percentage diagram (Figure 25) that conforms to the regional stratigraphic pattern found at other sites: a birch zone from 11,600 to 8,000 BP, subdivided by Ovenden (1981, 1982) into a lower Cyperaceae subzone and an upper Ericales subzone; a birch-spruce zone from 8,000 to 6,700 BP; and a birch-spruce-alder zone from 6,700 to the present.

Fifty samples of peat were analysed for plant macrofossils, and roughly 2,500 specimens belonging to 26 taxa were identified. These are summarized diagrammatically and grouped into seven zones, designated PM-1 to PM-7 (Figure 26).

During PM-1 and PM-2, from 11,600 to 10,900 BP, the site was under shallow lake water at the edge of the large, but diminishing Glacial Lake Kutchin, and the local vegetation consisted of abundant *Chara*, with frequent occurrences of *Potamogeton*, *Eleocharis*, and *Hippuris* growing in clear, neutral or slightly alkaline water. A steady increase in organic silt, a rise and peak of grass pollen, and decreases in *Chara* and ostracods suggest a transition from pond to marsh vegetation over a period of roughly 300–400 yr. PM-3, at 10,900–10,200, signalled a lowering of the water table and a transition from a grass marsh to a sedge fen, with increases in *Potentilla palustris*, *Cicuta mackenziana*, *Carex rostrata*, and *C. chordorrhiza*. By the end of this zone, at about 10,200 BP, dwarf birch and willow were members of the fen assemblage, and the grass *Glyceria*, now restricted to boreal areas south of 67°N, was common. Zone PM-4, from 10,200 to 9,600 BP, is a transitional unit reflecting a relatively rapid change from a minerotrophic fen to an ombrotrophic *Sphagnum* bog, with a short-lived peak of dwarf birch and the first colonizers among the heaths. Zone PM-5 spans 9,600–5,700 BP and represents the rapid development of this *Sphagnum* bog, with hummock-forming *S. fuscum* and *S. girgensohnii* the chief species, in association with *Ledum decumbens*, *Andromeda polifolia*, and *Oxycoccus microcarpus*, and *Sphagnum balticum* and *S. compactum* later becoming dominant. *Picea* pollen increases at 8,000 BP, indicating the spread of spruce woodlands on the surrounding uplands, but spruce did not occur in the mire communities of the lowland until later, in the subsequent zone, PM-6, where *Sphagnum fuscum* (briefly), *S. balticum*, and *S. compactum* were the chief mosses at the site. Finally zone PM-7, from 3,200 BP to the present, represented by highly humified peat with abundant *Polytrichum*, *Aulacomnium*, *Ledum decumbens*, and *Vaccinium vitis-idaea*, is interpreted as a heath-like community that developed as the polygon surface dried out locally.

Ovenden (1982) correctly takes a cautious position about possible vegetation reconstructions for the surrounding uplands, based on the Polybog pollen diagram. A tentative suggestion is that spruce woodlands became established on the uplands at about 8,000 BP, that black spruce expanded onto mire surfaces about 5,000 BP, by which time the modern configuration of communities had become established. The late glacial and early Holocene record is more difficult to

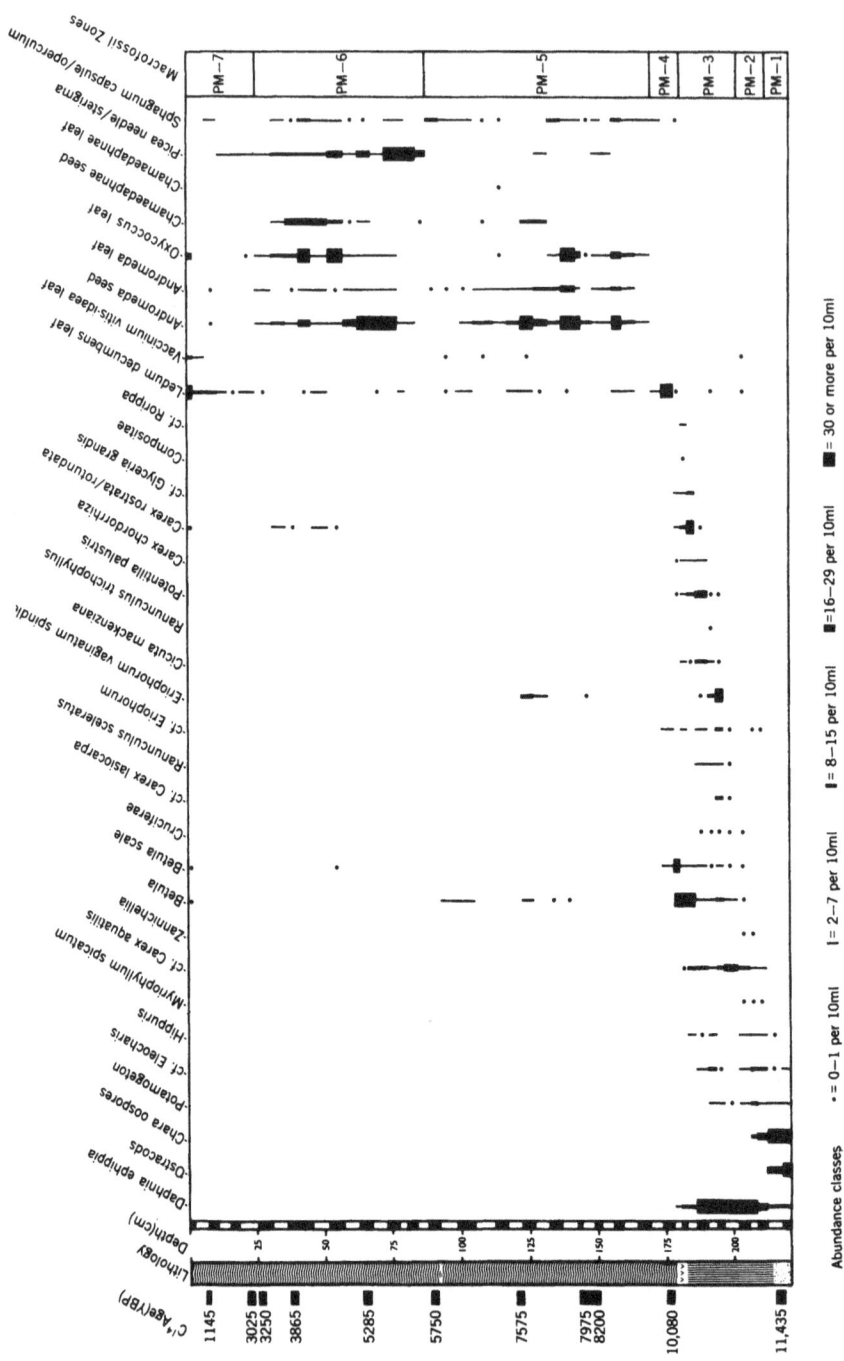

Figure 26 Diagram of macrofossil concentrations plotted against depth in sediment, after Ovenden (1982)

Abundance classes   • = 0–1 per 10ml   I = 2–7 per 10ml   I = 8–15 per 10ml   ■ =16–29 per 10ml   ■ = 30 or more per 10ml reconstruct beyond the simple suggestion that a shrub tundra of birch and willow, with abundant herb elements and slow increases in ericads, might have prevailed on upland sites.

This elegant investigation is the first adequately documented demonstration (in North America) of the sequence of plant communities that colonized an aquatic environment in a subarctic region, based on peat, macrofossil, and pollen analysis of a securely radiocarbon-dated section. It also illustrates clearly that although routine pollen analysis of mire deposits at high latitudes provides a very imperfect record of vegetation history, detailed macrofossil analysis can yield a detailed record. We will examine the palaeoenvironmental implications of the Polybog results in the next chapter.

(d) *Bluefish Caves*

The last site we shall consider is a group of three small caves at 67°09′N, 140°45′W, in an escarpment of limestone overlooking the Bluefish River (Figure 12). The caves contain a deposit of late glacial loess from which rich palaeontological and palaeobotanical material, and some cultural evidence, have been recovered (Cinq-Mars 1979). Pollen analysis of two sections of the sediments in cave II has been completed (Cwynar in Cinq-Mars 1979; Ritchie et al. 1982), and the results will be reviewed briefly here.

The sediments consist of a loess unit, approximately 0.5 m in depth, overlying the limestone bedrock, under a cap of fibrous, stoney moss humus also about 0.5 m deep. The loess unit has yielded a large number of vertebrate bone fossils, and two radiocarbon age analyses of collagen from two specimens provide an estimate of the upper and lower age limits of this unit: 12,900 and 15,500 BP (Cinq-Mars, personal communication 1981). The vertebrate remains are of horse, mammoth, dall sheep, caribou, bison, wapiti or elk, muskox, and several species of small mammals. The pollen spectra from the loess unit are similar to those from the late glacial sediments at Lateral Pond in that they are dominated by herb taxa and show a trend to high birch frequencies in the younger samples. The herb zone includes some of the same arctic-montane taxa characteristic of the Lateral Pond and Hanging Lake zones of the same 14-C age, namely, *Aconitum, Hedysarum, Pedicularis, Phlox, Polygonum alaskanum*, and *Oxytropis*. The pollen record of the upper sedimentary unit is poorly differentiated into subzones, not unexpectedly since the humus contains abundant evidence of disturbance: intrusive roots, frost-shattered pebbles, and a general absence of well-defined stratigraphy. However, the pollen spectra are dominated by spruce and alder with low frequencies of herbs, and in that respect they correlate well with the Holocene records from other sites in the North Yukon and adjacent Alaska (Ager 1975).

Pollen analysis of cave sediments is fraught with uncertainties, as the extensive literature on southern European sites illustrates (cf. de Lumley et al. 1973). Pollen and its containing matrix can be deposited by several agencies: wind, water, gravity, animals, and humans. Poor preservation, disturbance by frost action and other factors, and contamination of older levels by younger material are all common sources of error in pollen data from cave sediments. The Bluefish diagram illustrates these problems. Although the broad separation into a late glacial zone and a Holocene zone is clear, fine resolution into subzones would be unreliable. The suggested reconstruction for the late glacial vegetation at this site is a sparse herb tundra on upland surfaces and a sedge-grass-willow community in the valley bottom habitats. We will develop further the general ecological implications of such a reconstruction in Chapter 7.

3.5. SUMMARY OF THE PLEISTOCENE AND HOLOCENE VEGETATION

An attempt to summarize the history of the vegetation of the area from the beginning of the Quaternary to the present can be made if we supplement the factual record with hypothetical reconstructions to fill in the several large gaps. Reconstructions are made for significant time segments and, based on much subjective inference, the plant cover is suggested for the three main physiographic units that prevail: mountains; pediment surfaces, rolling plateaus, and glaciated uplands; and lowlands (Table 8). The reader should note that the radiocarbon chronology I use is based on my subjective estimations of average ages for the various biostratigraphic boundaries. There is considerable variation in dates between adjacent sites because of the large set of factors that can cause divergences from the true age of samples. This uncertainty explains the tentative chronology in Table 8, and the use of such phrases as 'at about 9,500 *Juniperus*...' As will be shown in the next chapter, I use the arithmetic mean of a number of age estimates of the same phenomenon – e.g. the early Holocene spruce rise.

The only evidence from the early Quaternary is the pollen assemblage from the lower sediments of the Bluefish Basin exposed along the Porcupine River, and they suggest a coniferous forest physiognomically similar to the modern southern boreal forest of western Canada, dominated by spruce, pine, and tree birch with associated willow, alder, and hazel. There is no palaeobotanical evidence for the middle Quaternary time span – i.e. from about 1 my to 100,000 yr – so I hypothesize that the frequent oscillations from glacial to interglacial climate indicated from evidence elsewhere in the Northern Hemisphere – chiefly the marine sedimentary record – controlled alternating cycles of tundra and boreal forest vegetation on most surfaces except upper montane habitats where tundra communities persisted continuously, changing only in extent.

Between 100,000 and 30,000 yr there is evidence from the unglaciated western portion of our area that at least one cycle of tundra to coniferous forest to tundra vegetation occurred, but it is likely that high-elevation sites retained some tundra throughout such a cycle while the major shifts from treelessness to forest occurred on intermediate and lower slopes. From roughly the beginning of the latest glacial cycle to the present the record is both continuous and more detailed so that regional and temporal distinctions can be detected.

During this glacial maximum when, if we accept the latest version of events (Hughes et al. 1981), Laurentide ice covered all of the eastern part of the area except possibly the distal portion of the Tuktoyaktuk Peninsula (Figure 5), the vegetation cover of upland surfaces was a sparse herb tundra while high-elevation montane habitats were largely devoid of plant cover and lowland habitats with impeded drainage supported grass-sedge marshes and willow carr. From about 14,000 to 11,000, as Laurentide ice retreated from its position centred on the area of the modern Mackenzie Delta (Figure 6), relatively rapid changes in vegetation occurred with considerable regional differentiation probably related to proximity to both the Beaufort Sea and the retreating ice sheets, and to such physiographic and edaphic variables as the calcium carbonate content of the rock types. In general most non-calcareous upland soils were occupied initially by a dwarf-birch shrub tundra. The dwarf-birch shrub tundra replaced the existing herb tundras on middle and lower slopes to the west of the Richardson Mountains, and was the first stable vegetation on newly exposed land surfaces on the Tuktoyaktuk Peninsula, east of the Mackenzie Delta. Xeric habitats, such as kame summits and ridge crests, supported *Shepherdia canadensis*–herb communities similar to localized facies that occur today on well-drained exposed bluffs with no immediate tree canopy (Plate 22). On calcareous surfaces everywhere, usually where dolomitic or limestone bedrock is at or near the soil surface, floristically varied herb tundras persisted or, in the case of eastern uplands freshly exposed by the disintegration of Laurentice ice, formed the initial cover. It is interesting to note that there is no evidence at this time, in the admittedly small number of sites, for any latitudinal zonation of vegetation.

At about 11,000 BP major vegetational changes occurred on most upland sites. In the North Yukon, north of the Old Crow Flats occupied then by the diminishing Glacial Lake Kutchin, there was an expansion of ericad genera to form the dominant dwarf birch-ericad heath tundra that prevails today. In poorly drained lowland tracts *Eriophorum-Carex*-ericad communities similar to those of the modern tussock tundras became established. *Populus balsamifera* groves and *Salix* thickets colonized many lakeshore and stream-side habitats, extending as far north as the coast as indicated by the wood found near Sabine Point (Rampton 1982, GSC-2022, age 9,940 ± 90). Similar changes occurred in southern and eastern regions, with the earlier dwarf birch *Shepherdia* shrub tundra being partly replaced by poplar groves and willow thickets, particularly on alluvium and lower slopes. At about 9,500 *Juniperus communis* expanded rapidly onto upland habitats over non-calcareous bedrock, sharing dominance with ericads and dwarf birch. Finally, at about 9,000 *Picea glauca* forests spread rapidly onto all upland surfaces as far north as about 69°30′ to the east of the Mackenzie Delta and to a short distance north of their present limit in the western section immediately to the north of the Old Crow Flats. By this time the modern patterns of upland and montane tundra communities were in place, most notably the dwarf birch heath tundras of middle elevations on non-calcareous soils.

Table 8 A summary of the vegetation history

Time segment (yr BP)	UNGLACIATED WESTERN REGION			
	Mountains	*Pediments and rolling uplands*		*Lowlands*
100,000 to 30,000	cycles of: discontinuous fellfield tundra herb and shrub tundra discontinuous fellfield tundra	cycles of: shrub-heath tundras spruce-tree birch closed boreal forest shrub-heath tundras		cycles of: sedge-grass-wi marsh spruce-larch-S *num* mires sedge-grass-wi marsh
24,000 to 15,000	fellfield tundra or barren	sparse herb tundra, discontinuous on xeric sites		sedge-grass-will marsh
14,000 to 11,000	s – dwarf willow fellfield tundras c – sparse herb tundra	s – dwarf birch–ericad closed tundra c – closed, rich herb tundra		continuous cott grass–sedge me
10,500 to 7,500	Differentiation of modern tundra types (as Chapter 6), varying spatially with rock type, aspect, slope, and stability; varying in time with climate but evidence is lacking	*Northern sites* s – expansion of closed dwarf shrub-heath tundras c – expansion of rich herb tundras to modern composition	*Southern sites* Rapid sequence of dwarf birch or herb tundra balsam poplar–willow–juniper spruce woodlands	black spruce–S, invasion of sou sites to form tre extension of co grass–sedge–er mires in the no
7,500 to present	no significant changes	little significant change except spread of alder into dwarf shrub tundras	modern patterns established by spread of tree birch and alder and slight retreat of spruce treeline	development o terned mires in and treed hum mires and palsa south

s = siliceous substrata; c = calcareous substrata

CIATED EASTERN REGION	
ng uplands	*Lowlands*
nentary, uncertain pollen evidence of coniferous-uous boreal forests	

ated)	(glaciated)
varf birch / *epherdia* shrub ch herb tundra	sedge-grass-willow marsh
nce of poplar-wil-iniper all associ-with dwarf birch; ed or modified by on of spruce wood-at 9,000; modern -tundra ished	development of cotton-grass–sedge–ericad mires in the north; black spruce–larch treed mires in the south; modern Mackenzie delta vegetation patterns initiated
n patterns formed ival of tree birch der; retreat of lati-al treeline to ly its modern posi-y 4,500	development of patterned mires in the north and of treed hummocky and palsa mires in the south

It will be clear to the reader from a comparison of the above summary and Table 8 with the descriptions of the modern vegetation in Chapter 4, itself a condensed account, that many fine details of plant community history are lacking. Some of these are provided in the body of this chapter with reference to particular sites, and the reader can find additional information in some of the original cited publications. But we must acknowledge that the existing network of sites is still sparse, and the resolving power of pollen analysis as a tool to reconstruct vegetation, as opposed to detecting biostratigraphic change, is limited. However, we can anticipate progress in the next several years in the light of refinements of site selection and in the application of new techniques for establishing sound statistical relationships between modern vegetation data and modern pollen values (see the articles by Prentice 1982, Walker 1982a, among others).

I have deliberately omitted from this chapter any discussion of problems of the origin and migration of the major taxa involved in these vegetation changes, and of questions of controlling or associated changes in environment. It is probably advisable to separate the record of vegetation change as registered in the pollen and macrofossil data from any palaeoecological inference. Environmental reconstructions based on these data are difficult and sometimes contentious, largely because the necessary quantitative relationships between measures of growth and reproduction of the important plants and the various factors of modern environment are poorly known, with a few exceptions. As a result circularity of reasoning or simply purely subjective assertion become likely if clear distinctions are not maintained between the record of vegetation change and the resulting palaeoenvironmental inferences. Chapter 6 is devoted to the latter, and to the related questions of species migration.

6
Palaeoenvironmental Reconstruction

One conclusion from Chapter 5 is that, so far, only the period from roughly 25,000 years ago to the present has yielded a sufficiently detailed fossil record to provide an adequate reconstruction of past vegetation. Three salient features of that record emerge, and in this chapter I will examine various hypotheses to explain them. The three major vegetational episodes or phenomena are:

1. *The full- and late-glacial (25,000–11,000) vegetation*, represented by predominantly herbaceous arctic-alpine taxa yielding relatively low pollen influx values.

2. *The early Holocene (11,000–7,000) period of rapid change in vegetation* – in the northern, unglaciated Yukon an abrupt transition from herb tundras to dwarf-shrub tundras; in the southern sector of our region, a dramatic change from tundra to a forested period, itself characterized in some regions by rapid changes in composition of the dominant species. The beginning of this period (~11,000) is signalled by increases in pollen influx over late-glacial values of at least an order of magnitude.

3. *The late Holocene (7,000 to present), a period that began with an abrupt increase in alder pollen followed by apparent stability of the regional vegetation patterns.*

A procedural thread runs through this chapter, in the interests both of minimizing the influence of my subjective, intuitive inclinations and of permitting the reader to keep clearly in mind the basic phenomena and evaluate for himself / herself the alternative hypotheses. For each of the above three major phenomena we will set out alternative hypotheses that fall into three general classes: (a) climatic, (b) other environmental factors, (c) ecological characteristics of the plants involved. We will examine the relevant evidence, including any sets of independent data that might be available, and where possible I will suggest which explanation appears to be valid or most likely. At the end of the chapter I will draw together whatever general palaeoenvironmental conclusions have emerged, but of course with this format the reader will be able to do the same in the light of whatever different insights he / she brings to bear on the material.

At this point we should remind ourselves of the outline of the state of general understanding of the past environments of the Northern Hemipshere during the Quaternary, a topic we touched on in Chapter 1. We know from the isotope record of deep sea sediments (Shackleton and Opdyke 1973) and related data from terrestrial sites (Kukla 1981) that throughout the Quaternary period the Northern Hemisphere environment has undergone roughly 20 oscillations from conditions promoting the growth and maintenance of continental and montane glaciers to conditions causing their disappearance from many regions. The time span for which our area provides data, the past 150,000 yr, encompassed in the Northern Hemisphere an interglacial of roughly 10,000–15,000 yr, between 127,000 and 115,000 yr, a long glacial with several minor oscillations (interstadials) between 115,000 and 10,000, and an abrupt termination of the glacial at 10,000 when the present interglacial began. The current consensus among climatologists, recently summarized by Berger (1981), is that the climate has varied in response to long-term variations in the earth's orbital relations, conforming to the basic astronomical theory elaborated by Milankovitch over 50 years ago. However, we should note that 'these global-scale changes in climate forcing will rarely, if ever, be manifested by similar climatic responses in all areas, due to the wave structure of the westerlies and the nature of the meridional circulations' (Barry 1982).

A final preliminary comment: may I caution readers not to accept too precisely the radiocarbon age limits I set on events or episodes. As we will discover when we examine such events as the arrival times of various taxa in the region, there are often large differences in radiocarbon age of a single event between adjacent sites. As Birks and Birks (1980, p 15) point out, radiocarbon ages of a palaeoecological event provide only an approximate estimate of its simultaneity at different sites, because of the many sources of error and discrepancy in the method. Readers can pursue this topic by reading the excellent review by Olsson (1974).

The chapter is organized into three sections, corresponding to the three major vegetational episodes or phenomena noted above, numbered 1, 2, and 3. Hypotheses developed to account for each will be numbered 1a, 1b, 2a, 2b, etc.

1. The Full- and Late-Glacial Vegetation

During the period >20,000 to ~11,000 total pollen influx is low at those sites with detailed and reliable records. The floristic composition of these assemblages is predominantly arctic-alpine. The vegetation of upland sites is reconstructed in Chapter 5 as a sparse, discontinuous tundra, predominantly a herb tundra in full-glacial time and later a herb-willow tundra with sparse dwarf birch occurrence. Lowlands supported continuous grass-sedge marshes with local willow thickets.

Hypothesis 1a. A cold, dry arctic climate analogous to the modern mid-arctic climatic regimes of North Banks Island and North Victoria Island prevailed in the

Table 9 A summary of selected climatic values based on the records published in the Climatological Summaries for Inuvik, Tuktoyaktuk, Lady Franklin Point, Holman and Sachs Harbour, Environment Canada, Atmospheric Environmental Service. I have derived the values for North Victoria by extrapolation.

	Inuvik	Tuktoyaktuk / Lady Franklin Point	Holman / Sachs Harbour	North Victoria Island
Mean daily July temperature (°C)	13	10	6	2
Mean daily August temperature (°C)	10	7	5	2
Mean date when daily mean temperature drops below 0°C	10 Sept	5 Sept	30 August	20 August
Mean annual total degree-days above 5.5°C (growing)	550	350	200	100
Mean annual total of degree-days above 0°C	1,000	950	550	500
Mean annual days with frost	300	325	350	>350
Mean date of first frost	20 August	15 August	20 July	10 July
Mean date lakes clear of ice	20 June	30 June	10 July	15 July
Mean date lakes freeze over	15 Oct	1 Oct	30 Sept	15 Sept
Annual mean precipitation (cm)	25	13	12	10
Annual mean snow (cm)	190	65	60	< 60
Annual mean daily temperature (°C)	−9	−12	−14	−15
% arctic air masses in July	60	80	90	95
% Pacific air masses in July	30–40	20	10	10

full-glacial, moderating slowly to an arctic climate analogous to the low-arctic climates that prevail today on South Banks Island and South Victoria Island. Data defining these climates are summarized in Table 9, but I stress that this hypothesis does not imply that one can reconstruct the full- and late-glacial climates of ice-free northwest Canada simply by displacing southwards the modern mid- and low-arctic regimes. It is likely that the substantially different conditions of land-sea-ice relationships during full-glacial conditions, when extensive areas of the Bering continental shelf were exposed by lowered sea level and when continental and montane ice masses existed to varying extents on both sides of the Bering Strait (Hopkins 1982), would have produced a climatic regime both in itself regionally variable and different from any modern high-latitude regime. None the less, the most parsimonious explanation rests on the observations that modern low influx values in the range 100–400 grains / cm^2 are restricted to mid-arctic stations (Ritchie and Lichti-Federovich 1967; Birks and Birks 1980), and that the modern northern limit of the range of dwarf birch (Porsild and Cody 1980, Figure 445) is coincident with the boundary between mid- and low-arctic western continental climatic regions (Maxwell 1980). The Hanging Lake influx data (Figure 21, modified slightly from Cwynar 1982) suggest that the climate between 20,000 and 16,000 was unfavourable for dwarf birch but that from 16,000 to 12,000 it exceeded marginally the tolerance threshold of that species. The modern vegetation analogue of the pollen assemblage recorded between 16,000 and 12,000, dominated by an association of dwarf willows and arctic-montane herbs with local occurrence of dwarf birch, is the chief vegetation type of modern upland tundras in South Banks Island (Zoltai et al. 1980) and South Victoria Island (Peterson et al. 1980).

Hypothesis 1b suggests that the low influx and predominance of herbs was due to the absence of dwarf birch from full-glacial landscapes, and the cause was not unfavourable climate but that it had not yet migrated into the northwest from its refugium south and / or southwest of the Laurentide ice cover. That explanation cannot be refuted at present, but it does appear to be unlikely. Dwarf birch influx values rise rapidly at all sites in both the glaciated eastern and unglaciated western regions at about 11,000, although some eastern sites show puzzlingly high values at 13,500 immediately following deglaciation. A migration route from southern Alberta would have been narrow and of very recent availability, so that migration across the 1,500 km to the lower Mackenzie and adjacent Yukon would have been accomplished in a few hundred years. It is more likely, but unproved, that dwarf birch survived the glacial in localized stations in the northwest, too rare and reproductively suppressed to be registered in the Hanging Lake full-glacial sediment.

A third alternative (*hypothesis 1c*) is suggested by Guthrie (1968, 1982), that the low pollen influx was caused by widespread grazing by large herbivores. There is evidence from northern Ellesmere Island (Svoboda 1983, personal communication) that local concentrations of muskoxen maintain patches of herb tundra in a closely clipped condition throughout the growing season, thereby inhibiting flowering and pollen production. The hypothesis cannot be refuted but it lacks entirely supporting data – as Hamilton (1982, p 714) points out, 'carbonaceous tundra and grassland paleosols are generally absent during this interval [in central Alaska] and radiocarbon-dated vertebrate fossils are scarce.'

Hypothesis 1a seems reasonable in the context of plant ecology, but it is not supported by a set of independent biostratigraphic data from the same region. Delorme et al. (1977) interpret their results of freshwater ostracod analysis from sediments with a radiocarbon age of 14,410, collected from a site near the Peel River (Figure 12), in terms of a climate very much warmer than the present, with mean annual temperature between 0.8° and 1.2°C, compared to the modern value of −10°C. Their interpretation appears to depend heavily on inadequately explained calibration functions derived from modern data on climate and ostracod distribution. It is also possible that the date is spuriously old; the authors state that they reject other dates from the same site because of possible marl contamination, and I suggest below, in discussing the poplar pollen peak they found in the same samples, that a younger age (10,000) is biostratigraphically more concordant with other data sets. Mackay (1978) refutes their interpretation by pointing out that the substantial late-glacial warming they infer would have left a record of massive thawing of ground ice in the coastal sediments but that no such evidence has been found. On the contrary, he points out that the sediments provide evidence that mean annual air temperatures remained several degrees below zero throughout full-glacial, late-glacial, and Holocene times. Further applications of the large set of modern ostracod-environment data would be of great interest and might help explain these conflicting hypotheses.

Until other evidence comes to hand to refute it, I accept hypothesis 1a, but suggest that the actual values cited in Table 9 for supposed analogues merely give an approximate estimate of the climatic conditions that might have prevailed. We will postpone consideration of other hypotheses about full-glacial Beringian climates until Chapter 7, because they are based on palaeontological data.

2. The Early Holocene Period

Several qualitative and large quantitative changes in the pollen and macrofossil records in the interval 11,000–7000 yr are registered at the sites examined in Chapter 5, summarized as follows: (1) Total pollen influx increases by at least an order of magnitude to values at several sites that are maximal for the entire Holocene. (2) Influxes of individual taxa (*Picea*, *Betula*, *Populus*, *Juniperus*, ericads) show rapid increases during this period from zero or low values to their maximum values for the Holocene. (3) At sites in the modern forest zone, some taxa (*Picea*, *Populus*, *Salix*, *Juniperus*) reach their maximal influx values at different times during the first few millennia of the Holocene. (4) The spruce pollen and macrofossil records, particularly from sites near and beyond the modern treeline on the Tuktoyaktuk Peninsula, suggest that the northern limit of continuous boreal forest advanced to the Beaufort Sea coast during this period and remained there for at least three millennia. (5) *Typha* (the cat-tail rush), though rarely abundant enough in samples to produce continuous percentage or influx curves, is more or less confined to this time span but is absent from the modern flora and pollen rain. (6) Dated macrofossils of poplar and larch suggest that both trees occurred in this region during the early Holocene well to the north of their modern limits. (7) The detailed record from Hanging Lake, beyond the modern treeline, shows a major increase in dwarf birch influx associated with marked increases in ericad, sedge, and *Sphagnum* influxes, interpreted as a major vegetation change from sparse herb tundras to the modern tundra communities. (8) Several sites show maxima of *Myrica* pollen, indicating a northern extension and subsequent regression of its range in the northwest.

The following seem to me the hypotheses most likely to account for the above phenomena. Some are alternatives or mutually exclusive, and I will indicate which appear to be the most securely supported; others overlap to some degree or have been developed as a corollary of another to account for a particular phenomenon.

Hypothesis 2a – a rapid warming of climate to maximal values for the Holocene. This explanation suggests that the total influx increases at about 11,000 can be explained by a rapid warming of climate to temperatures higher than at present, reflecting closely the estimated increase by 7% in global radiation of the Northern Hemisphere implicit in the Milankovitch orbital variation theory of climate change (Berger 1981). In particular, the extension of a continuous spruce forest to the northern coast of the Tuktoyaktuk Peninsula suggests that climate was more favourable than at present by the following increases over modern values: mean July temperature by 3°C, mean annual total degree-days above 5.5°C by 200, mean annual precipitation by 12 cm, mean annual temperature by 3°C. These values can be thought of as minimal since they are derived from the modern differences between Inuvik and Tuktoyaktuk, although there is some evidence (see below) that the early Holocene forests on the Tuktoyaktuk Peninsula had slightly better growth values than have modern woodlands at Inuvik.

Hypothesis 2b – climate change as in 2a combined with differential migration of certain taxa. This explanation proposes that neither climate change nor migration histories are alone adequate determinants of these changes, but that both operated, probably differently for different taxa. For example, a climatic warming is necessary to explain the replacement of tundra by forest in most southern and central parts of our area, but the successive rather than contemporaneous increases of poplar and spruce might have been due to differences in their migration rates. A variant of this hypothesis, proposed by Murray (1978) and adopted by Hopkins et al. (1981), is that balsam poplar survived in Beringia throughout the latest glacial and then spread rapidly in the late-glacial in response to abrupt warming and efficient dispersal mechanisms.

Hypothesis 2c – an abrupt warming as in 2a, followed by smaller-scale climate change. This is a corollary of 2a and it is proposed to account for the changes in pollen diagrams in the eastern, forested section, where the early Holocene is characterized by a rapid succession of tree and shrub taxa (birch, willow, poplar, juniper, and spruce). This explanation suggests that relatively minor changes in climate, with respect to the seasonality of precipitation for example, or indirectly to the progressive aggradation of permafrost, might have controlled the sequence of changes in the dominant taxa because of slight differences in their ecological amplitudes. Such an explanation implies that the early Holocene climates have no exact modern equivalent, because there is little evidence that these taxa are strongly differentiated at present in terms of climate. Such a proposition is compatible with theories of the orbital forcing of high-latitude climate; they indicate that the summer insolation maxima (7% greater than present-day) at high latitudes (60–70°) at 10,000 yr BP were accompanied by winter minima (Berger 1981). In other words, changes in the seasonal distribution of heat and precipitation during the period 11,000–7,000 might have determined the apparent vegetation changes recorded in many of the pollen diagrams from our area.

Hypothesis 2d – an abrupt warming as in 2a, followed by a period of climatic stability with the vegetation changes, particularly in the forest zone, being due to *species interactions* such as competition and resulting succession. This explanation is a direct alternative to 2c above.

Hypothesis 2e – the influx and other changes can be explained *independently of climate change*, being due to the *differential immigration of species* with intrinsically higher pollen productivities than the earlier tundra taxa, particularly the arrival of species of spruce, poplar, tree birch, and boreal species of willow; the absence of these species from the northwest for several millennia, after the region became ice-free, was caused by variable migration lag responses along corridors or routes from full-glacial refugia.

First, we will examine the main taxa separately in evaluating the above hypotheses, after which we will consider aspects of community dynamics.

2.1. PICEA

It is only recently, following the initial work by Birks and Peglar (1980), that attempts have been made to use pollen size statistics to identify the two species of *Picea*. Spear (1983) has pioneered this approach in our area, and Brubaker et al. (1983) in Alaska. Their preliminary results are in agreement, showing that *P. glauca* was the first to arrive and spread, followed by *P. mariana* later in the Holocene (Figure 14), but data are too sparse at present to enable us to generalize. We can explore both the indicator-species approach and evaluate data of first arrival of the genus at sites in the northwest to test the hypotheses (2a, 2b, and 2c). The autecology of *P. mariana* has been studied in detail by Alan Black at sites in the lower Mackenzie Valley, reported in Black and Bliss (1978, 1980). They give minimum average monthly temperatures for June, July, and August of 11°, 14°, and 11°C for adequate reproduction, pointing out that these are 'estimated from ~40 km south of modern forest line and suggest forest line is out of equilibrium with modern climate.' They suggest that conservative climatic reconstructions could be based on these values, which were derived from germination and establishment experiments in both the field and laboratory. Their evidence supports Nichols's (1976) original suggestion that the present treeline is not in equilibrium with climate by noting the failure of forest regeneration 'after a fire occurred in stands originating prior to 1850.' When it is reproducing sexually, black spruce 'maintains a relatively constant seed population which is little affected by the burn interval, is not destroyed by fire, and accumulated on the tree over a period of years. *Larix laricina* and *Picea glauca* on the contrary, release seed annually, maintain no such seed population, and therefore must seed from survivors over longer distances' (Black and Bliss 1980, p 351). Payette and Gagnon (1979) and in more detail Gagnon (1982) have shown that black spruce can persist as prostrate krummholz, reproducing vegetatively, beyond the climatic limit for seed reproduction.

Unfortunately we have no such helpful published work on the autecology of *P. glauca* in our area, but we can draw on observations on populations in Alaska (Zasada 1980; Zasada and Gregory 1969; Zasada and Lovig 1983; Viereck and Schandelmeier 1980) and elsewhere in North America (Elliott 1979; Larsen 1980, Rowe, 1955, Sutton 1969). Its range limit in the northwest coincides closely with that of *P. mariana*, so that there are general correlations between its distribution and various macroclimatic values similar to those of black spruce, but its edaphic preferences and reproductive biology undoubtedly imply that the micro-environmental factors that operate in determining its dispersal, seed germination, establishment, growth, and reproduction are quite different, though incompletely known. White spruce populations in interior Alaska show great year-to-year variation in seed production, with frequent years of little or no yield (Zasada 1970). Seed is shed annually, in fall, and dispersal over frozen snow and ice surfaces is common. Seed germination occurs normally in early summer (June) and is sensitive to lower than normal temperatures. Seedling establishment requires an open, unshaded predominantly mineral substratum with little or no competing vegetation and moderate to high soil nutrient levels. This edaphic tolerance appears to control its habitat occurrence at the western treeline, where it is confined to either well to moderately drained upland mineral soils or seasonally flooded alluvium. It forms treeline in rolling, well-drained terrain of our area, reproducing by layering (Plate 25) near its northern limit, and occurring as krummholz bushes in the dwarf shrub tundras beyond the limit of tree growth. Zasada and Gregory (1969) report that summer climatic conditions are critical in determining the initiation of cone buds – a warm, dry period in June and early July produces heavy seed crops the following year. These findings have been confirmed and elaborated by Owens and Molder (1977) and Owens et al. (1977). Garfinkel and Brubaker (1980) demonstrate that radial growth of white spruce is correlated with current summer temperature and with the previous autumn temperatures.

White spruce is less tolerant of high permafrost table levels than black spruce, with the result that where well-drained uplands and peaty lowland occur together, as they do throughout much of our area, both spruce species form treeline, the particular situation being determined by habitat type and land-form.

If we follow Black and Bliss's (1980) recommendation for climatic reconstruction, and if we assume that the demonstrated macroscale temperature limits for black spruce apply also to white spruce because of their coincident northern limits, then the influx and macrofossil data from the Tuktoyaktuk tundra sites (Figures 13 and 14) suggest strongly that the regional climate between about 11,000 and 8,000 changed rapidly, reaching values of at least 3°C for the July mean temperature and 200 degree-days per annum above 5°C, higher than modern values. The influx values do not by themselves answer the question: 'Was there a response lag in the pollen data due to migration time delay, or is the influx curve a direct registration of climate change?' We can test these alternative hypotheses (2a and 2e above) by plotting the radiocarbon ages of the threshold values of spruce frequency against geographical location and scrutinizing the data for the presence or absence of a trend. Establishing threshold values is not easy. Percentage values must be used because we have no reliable values for the modern influx of spruce across treeline. The spruce pollen frequency in modern lake sediment samples across the transition region from northern boreal forest to shrub tundra in the Inuvik-Tuktoyaktuk area drops from >20% to about 10% (Ritchie 1974), so the latter value might be an appropriate threshold value to indicate spruce presence. However, we should note carefully the sources of possible error in such an assumption. First, the percentage of spruce, or any other taxon, can vary in time without change of its actual pollen input for the familiar reason that variations in the actual input of some other taxon, especially a prolific pollen producer such as alder, will cause percentage changes in all other taxa. So 10% *Picea* pollen in a landscape with a scattering of alder bushes might mean different spruce represen-tation than 10% in samples from a landscape totally devoid of alder. Secondly, the modern pollen rain of an area contains a significant and variable component of spruce and other pollen derived from distant, background source areas, at present the vast boreal forest region to the south and southeast of our area. In full-, late-glacial, and early Holocene times, however, this landmass was occupied by Laurentide ice and we might expect that the concentration of long-distance transported background pollen rain would have been significantly less than at present. The exotic background input becomes more important in herb- and sedge-dominated tundras where the local total pollen production is less than in adjacent shrub tundra and boreal forest zones, and the long-distance element is proportionally larger (e.g. data in Ritchie and Lichti-Federovich 1967; Fredskild 1969). The effect can be seen in the data set from the Mackenzie Delta area, and in samples from Banks Island (unpublished data); the sedge tundra sites on the distal portion of the Tuktoyaktuk Peninsula (Ritchie 1974, Table 3), where dwarf birch is less abundant than in the shrub tundras to the immediate south and west and where alder is absent, have higher spruce values (mean 14%, range 12–18%) than sites in the shrub-heath tundra closer to spruce treeline.

In other words my choice of 10% as the threshold spruce value might not be an accurate measure of spruce presence at a site, and it is probably an overestimate. However, it will not affect the outcome of the basic question: 'Is there any geographical trend in the data?' In tabulating the 14-C age of the 10% spruce threshold I have extended the area beyond our immediate study region to include more distant sites in central Alaska and thus improve the chances of revealing a trend (Table 10, Figure 27). However, I have excluded western Alaska from consideration mainly because it has been adequately dealt with by Hopkins et al. (1981), and it is clear from their compilation and pollen diagrams from sites in western Alaska (Ager 1982) that spruce did not reach there before 5,000 yr BP, so we can eliminate western Alaska as a potential source area for the spruce populations that first reached our area.

Figure 27 and Table 10 give the radiocarbon ages of the 10% spruce threshold from 22 sites in northwest Canada and east-central Alaska; I include only sites which I consider to have adequate chronological control, primary lake or mire sediment, and no doubtful or ambiguous results. The ages range from 7,900 to 10,900 (mean 9,095, SD = 758), but there is no evidence of trend in any direction.

Table 10 Radiocarbon age of the 10% spruce pollen threshold at 22 sites in northwest Canada and Alaska. In cases where the 10% level does not coincide with a dated level, the age was determined by linear interpolation.

Site and reference	Latitude	Longitude	Age (yr BP)
Eildun Lake, Slater (1980)	63°08′	122°46′	8,500
Sweet Little Lake, Figure 19	67°38′	132°00′	9,100
Lateral Pond, Ritchie (1982)	65°57′	135°31′	9,000
Tyrrell Lake, Ritchie (1982)	66°03′	135°39′	8,800
M-Lake, Ritchie (1977)	68°16′	133°28′	8,700
Twin Tamarack Lake, Figure 16	68°18′	133°25′	8,200
Tuktoyaktuk-5, Ritchie and Hare (1971)	69°03′	133°27′	10,000
Sleet Lake, Spear (1983)	69°17′	133°35′	10,000
Eskimo Lake pingo, Hyvärinen and Ritchie (1975)	69°25′	131°40′	9,500
Hendrickson Island pingo, Hyvärinen and Ritchie (1975)	69°32′	133°35′	8,300
Richards Island pingo, Ritchie (1972)	69°26′	134°30′	9,300
Hanging Lake, Cwynar (1982)	68°23′	138°23′	10,500
Polybog, Old Crow, Ovenden (1981)	67°48′	139°48′	7,900
L Site, Old Crow, Ovenden (1981)	67°35′	139°32′	8,100
RU Site, Old Crow, Ovenden (1981)	67°27′	140°45′	9,000
CB Site, Old Crow, Ovenden (1981)	68°06′	140°56′	9,000
SBA Site, Old Crow, Ovenden (1981)	67°47′	139°50′	9,100
Birch Lake, core I, Ager (1975)	64°19′	146°40′	10,900
Birch Lake, core II, Ager (1975)	64°19′	146°40′	9,500
Lake George, Ager (1975)	63°47′	144°30′	9,200
Alatna Valley, Brubaker et al. (1983)	67°30′	153°30′	8,500
Antifreeze Pond, Rampton (1971)	62°21′	140°50′	9,500

It is likely that the variation in apparent age is due primarily to errors in the radiocarbon age estimates since there is as much difference between samples from the same site or closely adjacent sites as between samples from widely spaced sites. For example, the Birch Lake and Lake George sites give values differing by 1,700 yr; the five mire sites in the Old Crow region range from 7,900 to 9,100; and M-Lake and TT-Lake, only 1 km apart, differ by 500 yr.

This conclusion is at variance with the suggestion of Hopkins et al. (1981), which they point out is quite speculative, that spruce migrated from south-central Manitoba, across western Canada, reaching the Mackenzie Delta at 11,500 and farther west into Alaska by 8,000–9,000 yr, finally reaching western Alaska at 5,000 yr. Their suggested migration route (Hopkins et al. 1981, Figure 4) would bring spruce from south-central Manitoba to Great Slave Lake in about 1,400 yr, thence to the Mackenzie Delta in 1,000 yr, and involve range extension rates averaging about 1 km / yr. That rate is four times the value derived by Davis (1981) for postglacial spruce migration in eastern Canada. Also much of the proposed route would not have been ice-free until about 10,000 yr (Prest 1970). It is unfortunate that Hopkins et al. overlooked the key Lofty Lake and Alpen

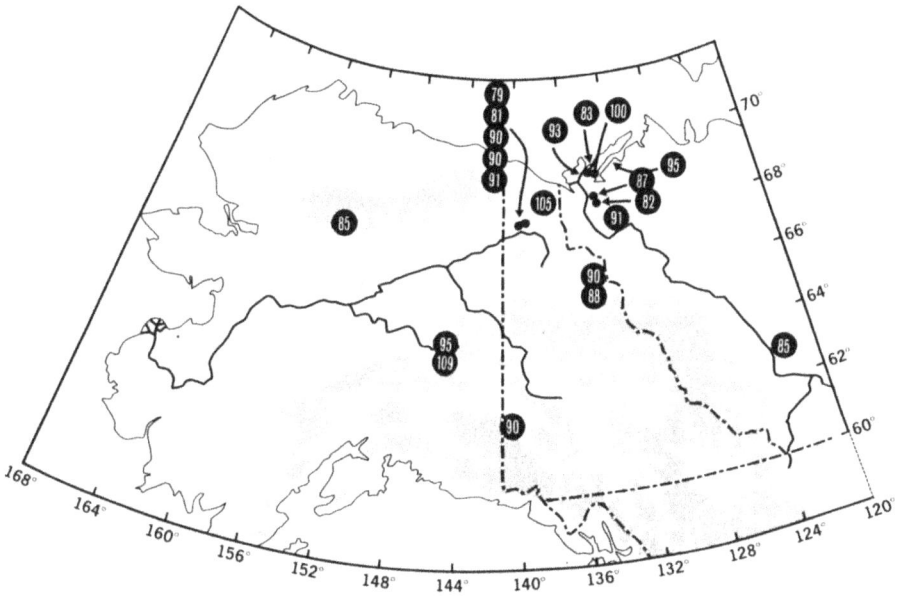

Figure 27 A sketch map of northwest Canada and Alaska. The circled numbers are radiocarbon age in hundreds of years of spruce arrival, as indicated by the 10 per cent pollen value. The sites are listed in Table 10.

Siding sites in central Alberta (Lichti-Federovich 1970, 1972). They provide a secure starting-point to test a hypothesis of spruce migration from a southern refugium to the northwest of the continent. If my earlier rejection of the Twin Lakes, Inuvik site (Chapter 5, p 97) is valid, and the 22 dates shown in Table 10 seem to confirm that conclusion, then we require a spruce range extension from central Alberta at 10,800 to the Mackenzie Delta via the foothills corridor area and the Mackenzie River valley, or to the Tanana valley (Alaska) and Interior Alaska via the lowland passes of the Tintina Trench as suggested by Cwynar (1982), by the mean date of 9,050. Let us assume, because of the absence of apparent trend, that the variation in the threshold dates is a function of intrinsic variability in the radiocarbon estimates, for a variety of unknown but likely reasons, the chief of which is probably variation in sediment chemistry from site to site. Then we can reasonably accept the mean arrival date as the best estimate for the area as a whole. That is, spruce migrated the 1,800 km from Lofty Lake to the Mackenzie Delta in 1,800 radiocarbon years, and if we assume an even rate of migration we should expect arrival times at intermediate localities roughly according to Table 11. My colleague Glen MacDonald has in train a project to test the migration hypothesis by sampling the spruce rise at a series of sites between central Alberta and the Mackenzie Delta. Until his results are to hand we can only speculate that the implied migration rate, 1 km / yr, or four times the value

Table 11 Actual and predicted times of arrival of spruce (10% pollen threshold), assuming an even rate of migration from the Lofty Lake site, Alberta to the Mackenzie Delta region

Site	Actual arrival	Predicted arrival
Lofty Lake, Alberta	11,500	
Caribou Mts, Alberta		10,900
Horn Plateau, NWT		10,300
Norman Wells, NWT		9,700
Mackenzie Delta	9,100	

estimated by Davis (1981) from a very large data set in eastern North America, is remarkably high. None of the tree species she analysed gave rates greater than 400 m/yr.

The implication of such a migration rate, if we assume 10 yr as the average minimum age of reproductive maturity in spruce (Zasada and Gregory 1969; Fowells 1965; Wright 1964), is that the average seed dispersal distance from trees at the advancing front would be 10 km. *Picea glauca* sheds winged seeds annually in fall, and even if we assume wind dispersal over open terrain, sometimes frozen and snow-crusted, this is a rapid rate compared to the available modern estimates. Viereck (1975) gives seed dispersal distance of twice tree height, and Waldron (1965), Rowe (1955), and Zasada and Gregory (1969) give values from 60 to 300 m. Dobbs (1976) provides similar values for primary dispersal – that is, the distance the seed travels in flight from the parent tree – but points out that greater distances can occur when secondary dispersal occurs, as for example by wind over snow or by watercourses. It is possible that when spruce reached the upper Mackenzie River, where it flows out of Great Slave Lake, river transport of seeds, and possibly vegetative propagules such as whole trees or branch systems carried downstream in spring and other floods, might produce a greatly accelerated migration northwards.

We should note, however, that the process of plant spreading is likely to be far more complex than the above simple considerations imply. Carter and Prince (1981) suggest that epidemic threshold theorems provide a useful depiction of the process of invasion. They show that in addition to the efficiency of seed dispersal, the number and availability of suitable sites and the number and longevity of source plants are important variables influencing migration rate. The migration hypothesis remains neither refuted nor confirmed, merely unlikely.

The alternative explanation is that spruce survived the full-glacial in Beringia, an idea that was proposed by Hopkins (1972) and Matthews (1976) but later abandoned (Hopkins et al. 1981). The palaeobotanical evidence, such as Rampton's (1971) 27,000 yr spruce pollen percentage rise at 'Antifreeze Pond,' Cwynar's (1982) influx and percentage values at 21,600, and the Bonnet Plume (Hughes et al. 1981) macrofossil sample at 36,900, can be interpreted as evidence of spruce in the northwest during the interstadial preceding the latest glacial, but there is no palaeobotanical evidence for persistence throughout that glacial. However, the taxonomic differentiation of *Picea glauca* into a northwestern race, *P. glauca* var. *porsildii* (Raup 1947), and the recent chemotaxonomic demonstration by von Rudolf et al. (1981) that samples of this smooth-barked variety from North Yukon display a richer genepool than other populations and that the variable terpene patterns could be explained by a 'genetic make-up ... derived from those that survived the last glaciation in northern refugia,' provide indirect support for a survival hypothesis.

Raup and Argus (1982), apparently largely on the basis of presumed ecotypic differences in white spruce populations, 'believe it reasonable to suggest therefore that populations of spruce probably lived through the last glaciation in the eastern valleys and foothills of the mountains west of the Mackenzie River.' They postulate that 'when the Tuktoyaktuk Peninsula was free of ice with tundra established, this spruce would have been nearby and already adjusted for a forest-tundra habitat. It then began a long migration eastward and southeast-ward, forming the arctic timberline and the subarctic open park-like forest.' There is no palaeobotanical evidence to support these propositions. White spruce was present in southern Alberta just prior to 10,000 BP (MacDonald 1982), and elsewhere in the Canadian prairies at the same time (Ritchie 1976), so it would appear more reasonable to suggest a migration northeastward in the wake of the diminishing Laurentide ice sheet towards the modern treeline between Hudson Bay and the area northeast of Great Slave Lake than to require migration southeastward from a glacial refugium in the Richardson Mountains.

However, the question of the origin of northwest spruce populations remains open; further work, to enable compilation of detailed migration maps, and to analyse thoroughly the population genetics of white spruce, will clarify and perhaps answer it (Neinstaedt and Teich 1971).

If spruce was present throughout the full- and late-glacial, though not produc-ing enough pollen or macrofossils to be registered in primary sediments, then the rapid increases in influx and percentage at about 9,000 yr require a major climate warming, supporting hypothesis 2a, except that the delay in response between 11,000 and 9,000 is inexplicable. However, if the migration pattern suggested by the data (Table 11) proves to be valid then the increases can be adequately explained without climate change (hypothesis 2e).

The spruce pollen and macrofossil records from sites east of the Mackenzie Delta at the modern arctic treeline provide evidence that the northern limit of the continuous boreal forest advanced to the Beaufort Sea at ~9,000, remained there for at least three millennia, and had retreated to approximately its modern position by 5,000 yr BP.

Hypothesis 2a explains the extension north and subsequent contraction of the continuous boreal forest as a response to a period, between 9,000 and 5,000, when the climate was more favourable than at present by the following estimated *increases*; mean July temperature, 3°C; mean annual total degree-days above

5.5°C, 200; mean annual precipitation, 12 cm; mean annual temperature, 3°C. Note that we have already concluded that this climatic warming began at 11,500 and that it was very rapid, reaching maximum values by not less than 10,000. Therefore the lag in the spruce response represents the migration delay from southern refugia, if we accept our conclusion above that a Beringian refugium is unlikely.

No reasonable alternative hypotheses exist to explain the retreat of the spruce treeline at ∼5,000 BP, unless we are prepared to entertain the possibility that recurrent forest fires caused the late Holocene replacement of spruce woodlands by tundra. Nichols (1976) has provided convincing evidence that fire can determine the position of treeline and that the minor changes in time can be ascribed to fire rather than climate, but he also points out that the frequency of occurrence of natural fire is controlled by synoptic-scale climate events so that climate remains the primary operative factor.

The Sleet Lake spruce influx curve (Figure 14, from Spear 1983) is compelling evidence to support his conclusion that 'during the period ... 9,000 to 5,000 yr BP continuous forest grew at least as far north as Sleet Lake.' This detailed analysis confirms and elaborates the earlier palynological result from the nearby Tuktoyaktuk-5 site (Figure 13, Chapter 5), and we have discussed the vegetational implications in the previous chapter.

Spear (1983) has collated the spruce macrofossil evidence, reproduced here in Table 5, and it supports the pollen data. In particular, spruce stumps in growth position (Plate 28) on parent materials of a clearly non-transported nature provide secure evidence of past tree growth. The number of dated stump samples is too few for a detailed analysis of the increment growth pattern. However, two stumps have yielded ratios of bole diameter to ring totals (roughly equal to bole age) well outside the range of modern white spruce in the forest-tundra near Eskimo Lakes and between the ranges for trees on upland sites and those on delta alluvium near Inuvik (Figure 28). One of these stumps, that shown in Plate 28, has a radiocarbon age of 4,940 ± 140, and the other, collected by Spear (1983, Figure 2), an age of 6,620 ± 70 yr.

It is obvious that the sample sizes of such a comparison are totally inadequate for firm conclusions, but there is an interesting suggestion that if the fossil stumps were representative of a forest on mesic slopes, we could accept a reconstruction in terms of a continuous spruce forest with tree dimensions at least as large as those found in the Inuvik area on the most productive upland sites.

We should not overlook changes in coastline as a factor that might have influenced the effective summer warming responsible for the early Holocene spruce advance, for example at the Sleet Lake site. Mackay (1963) has reported that the coast of the Tuktoyaktuk Peninsula must have been several kilometres north of its modern position in the early Holocene; subsequent drowning and recession has been due to both eustatic rise in sea level and isostatic coastal

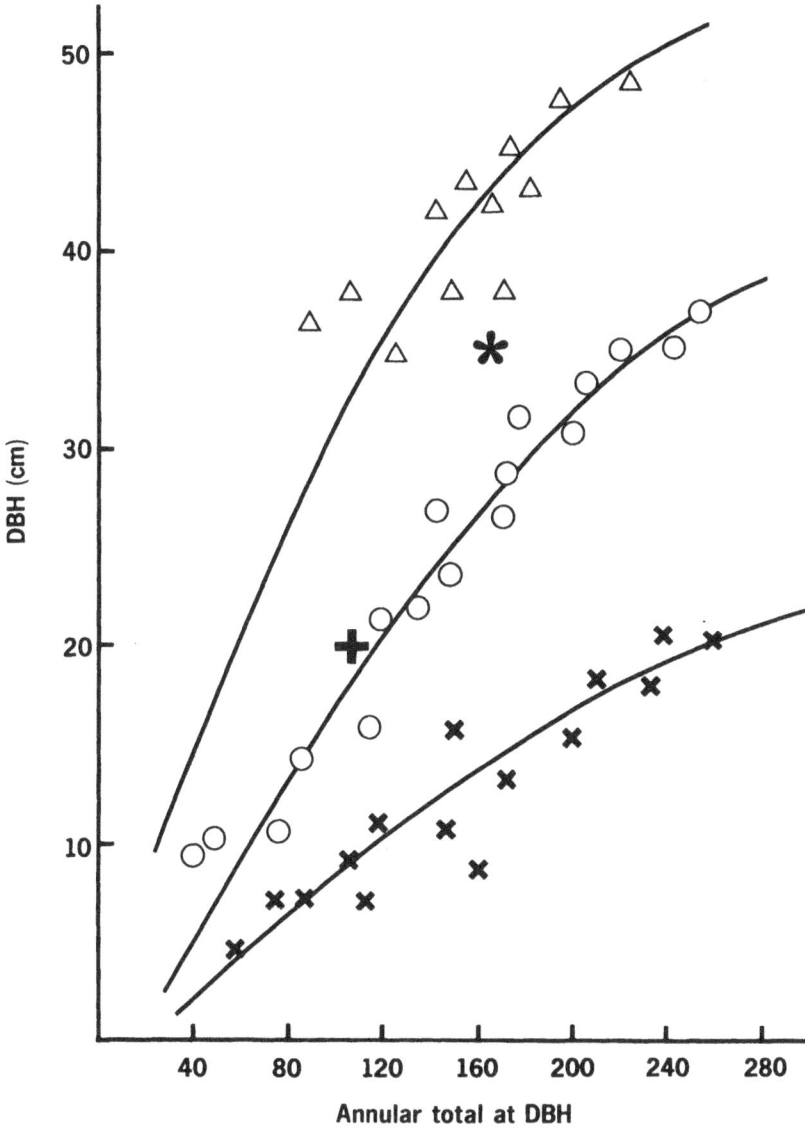

Figure 28 White spruce stump growth ring totals plotted against diameter at breast height, for modern samples on alluvium in the Mackenzie delta (△), on uplands near Inuvik (○), and on uplands at treeline (✕). + is the spruce stump at 69°27′ N, 132°10′ W; * is the spruce stump at 69°07′ N, 133°16′ W referred to in Table 5.

depression. It is impossible to estimate the possible effect that this might have had on the climate of the Tuktoyaktuk peninsula.

2.2. POPULUS

Hopkins et al. (1981) have assembled the macrofossil record for poplar in the northwest and conclude that it 'argues strongly for the persistence of cottonwood [*Populus balsamifera*] throughout the last cold period'; they cite as further support Murray's (1980) description of outlier, relict, clonal stands in northeast Alaska well beyond the modern treeline, apparently similar to those recorded from the Mackenzie area (Plate 20, Chapter 4). There are actually no records from the latest glacial: the record is blank between the 35,600-yr-old wood sample at 70°17′N, 148°31′W, and a sample in 13,700-yr-old peat at 69°42′N, 154°52′W, both in northern Alaska. However, as Hopkins et al. (1981) point out there is extensive evidence that balsam poplar was widespread in the northwest in the early Holocene. In our study area a sample of *Populus* wood, not identified to species, in an exposure on the North Yukon coast near Sabine Point, at 69°02′N, 137°38′W, was dated at 9,940 ± 90 (GSC-2022) and it is well beyond the modern limit of poplar. The matrix is described as a unit of 'mudflow debris overlying preglacial sediments,' and Rampton (1982) interprets the record as indicating a climate warmer than at present.

The pollen record for poplar from recent analyses of lake sediment cores from the northwest shows a distinct peak in percentage between 11,000 and 9,000 (Table 12). Only Brubaker et al. (1983) have identified sub-fossil pollen of the two species of poplar (*P. balsamifera*, *P. tremuloides*), and they find that the early poplar was *P. balsamifera*. I am making the assumption here that the poplar pollen at the other sites in Table 12 was from *P. balsamifera*. The data suggest that poplar woodlands expanded rapidly into the tundra landscapes of the northwest at about 11,000, reaching the north coasts by 10,000. However, the very low values at the northernmost lake sediment site lead Cwynar (1982) to suggest that these northern extensions were probably only scattered, local groves. Only the single sample reported by Delorme et al. (1977) at 451 cm in their site 2, with an age of 14,410 ± 110, indicates the presence of poplar earlier than about 11,500. This site presented dating problems; one 14-C determination was rejected by Delorme et al. (1977, p 2037) because of suspected marl contamination, and I suggest that the 14,410 date might also be too old. In any event it might be advisable to await confirmation from other sites before accepting that poplar was present in the area earlier than 11,500.

The plumed seeds of poplar are widely disseminated by wind, and balsam poplar is an early colonizer of immature, recently exposed or deposited mineral substrata, so rapid spread and establishment is feasible. However, in the absence of positive evidence for persistence in Beringia throughout the latest glacial it is not possible to solve the problem of the poplar record: Did it survive in the

Table 12 The occurrences of a poplar pollen percentage peak in pollen diagrams from the northwest, including one Alaskan site. Unless indicated, locations are on Figure 12

Site and location	Maximum %	Radiocarbon range
Alatna Valley, 67°30′, 153°30′ (Brubaker et al. 1983)	30	11,000–9,500
M-Lake, Inuvik (Ritchie 1977) (Figure 17 this volume)	35	11,000–9,500
TT-Lake, Inuvik (Figure 16)	15	10,500–9,000
SW-Lake (Figure 19)	10	10,500–9,000
Peel River, site 2 (Delorme et al. 1977)	32	14,410–9,890
Eildun Lake, 63°08′, 122°46′ (Slater 1980)	25	10,700–10,000
Hanging Lake (Cwynar 1982)	1	11,100–10,460

northwest, or did it migrate at remarkable speed from its ~12,000-yr-old sites in central Alberta (Lichti-Federovich 1970)? A further difficulty with the poplar pollen record is the familiar one of erratic preservation in sediments. For example, the SW-Lake diagram (Figure 19) shows the clear poplar peak in the lower samples, but the upper samples including those from the present are almost devoid of poplar pollen although both species make up a significant part (~40%) of the continuous boreal woodlands surrounding the site.

I conclude that the poplar evidence provides limited support (the dated log beyond modern treeline) for hypothesis 2a (climate warming). The pollen sites, all within the modern range of both poplars, suggest a warming to replace tundra with forest, but the timing remains uncertain in the absence of clear evidence for or against a migration lag.

2.3. LARIX

The larch pollen record is fragmentary because the pollen production is low and the large thin-walled grains appear to be overlooked in some investigations. It has been recorded at the following sites in our area: Eskimo peat bank (Mackay and Terasmae 1963) from at least 8,000 yr; M-Lake (Ritchie 1977) from 9,000 yr; Lateral Pond (Ritchie 1982) from about 9,000; Eildun Lake (Slater 1980) from about 9,000 yr.

These data suggest a time of arrival about 9,000, but otherwise contribute little to the discussion of the hypotheses in question. However, there is larch macrofossil evidence (Zoltai and Zalasky 1979; Zoltai 1977, personal communication) that provides some support for hypothesis 2a. Wood samples of larch, with radiocarbon ages of 7,510 ± 140 (BGS-472), were recovered from organic sediments at two nearby localities (69°05′N, 134°07′W and 69°05′N, 133°52′W) about 60 km beyond the present northern limit of larch. The modern distribution of *Larix* in the

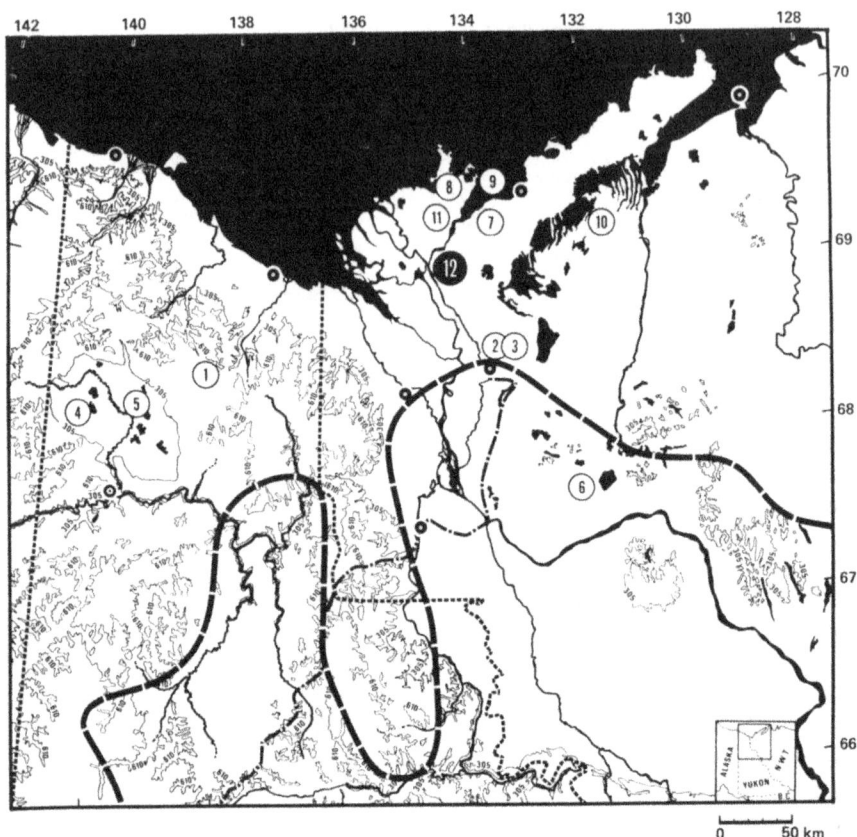

Figure 29 The numbered circles show the locations of the sites listed in Table 13 from which *Typha* pollen records have been abstracted. The solid circle numbered 12 is also the site of the *Larix* radiocarbon-dated macrofossils reported by Zoltai and Zalasky (1979) and referred to in the text. The broken heavy line shows the northern limit of *Larix*.

northwest, as Zoltai (1973) and Porsild and Cody (1980) have observed, forms an unusual pattern (Figure 29) showing no obvious correlation with macroclimate. Until a fuller sub-fossil record and a clearer understanding of the autecology of larch become available, its palaeoenvironmental value will remain limited.

However, the elegant study by Gagnon (1982) of larch macrofossil evidence near treeline in the Richmond Gulf area of Nouveau Québec demonstrates that larch can be a better indicator of palaeoclimate than spruce. Spruce can maintain its position in spite of climatic deterioration by persisting vegetatively, often as prostrate krummholz. Larch, in contrast, does not reproduce vegetatively by layering and is therefore dependent on reproduction by seed for its persistence and spread. Unfortunately the pollen record of larch in diagrams from our region is too sporadic to permit useful interpretation.

2.4. BETULA

Since the classical first investigations by Livingstone (1955a), dwarf birch pollen data have been important at all sites in the northwest of the continent in delimiting pollen zones and in generating hypotheses about vegetation and climate changes. A few preliminary points of clarification are needed. The taxonomy of the dwarf birches is difficult, and the resulting use of different names both in floras and manuals and by different field botanists can be confusing. Both *B. nana* and *B. glandulosa* have been identified in our area. The former is a variable circumpolar arctic-montane species; the latter is confined to North America and has a boreal–low-arctic range. I choose to follow Porsild and Cody (1980) and assume that all dwarf birch in the area is *B. glandulosa*, recognizing the point made by Hultén (1968) that considerable introgression has probably occurred between the two populations.

The microscopic distinction between the pollen of dwarf birch and tree birch (*B. papyrifera*) is based on size measurements, but there is some overlap between the ranges for the two species. The practice in our laboratory of using an arbitrary mean grain diameter of 20 μ, as measured in the equatorial plane, as the upper limit of dwarf birch has been corroborated by the macrofossil record, and of course is based on measurement of modern populations of both taxa (Ives 1977). Finally, it is important to note that dwarf birch is a prolific pollen producer; investigations of its present-day pollen representation at sites in our area show that plant communities with 2–3% dwarf birch ground cover give pollen percentages in adjacent lake sediments of 13–28% (Ritchie 1977, Table 6; 1982, Table 1). If this is not taken into account a false impression of the vegetational significance of dwarf birch pollen frequencies results. It is only recently as pollen concentration and influx values have been determined that an accurate chronology of the major birch increase has been demonstrated. Most percentage diagrams show a large percentage increase at about 14,000 yr, but the influx values show that this is an artefact of the frequency data and that the large increase in birch pollen input occurred at about 11,500. For example, influx data for dwarf birch at the Hanging Lake (Figure 21, modified from Cwynar 1982), Lateral Pond (Figure 23, modified from Ritchie 1982), M-Lake (Figure 17, modified from Ritchie 1977), and TT (Figure 16) sites all show increases from less than 200 to between 1,000 and 2,000 grains cm^{-2} yr^{-1} at 11,500, and the shapes of the influx curves from the three sites are identical, reaching maximum values at about 10,000 yr. I conclude that dwarf birch was present in very small amounts in the sparse, predominantly herbaceous tundras at these sites prior to 11,500, when it expanded rapidly. Brubaker et al. (1983) draw a similar conclusion from pollen diagrams of lake sites in the central Brooks Range of Alaska. Such evidence supports hypothesis 2a that there was a rapid warming and that climate change alone is an adequate explanation of these vegetational changes. Conversely the hypothesis of migration lag in birch is refuted since the influx data suggest that it was present in the area at least 16,000 yr ago. Unfortunately, these simple relationships break down when we examine the influx data from one of the two major sites on the Tuktoyaktuk Peninsula (Sleet Lake, Figure 14). There, birch influx is high ($\sim$3,000 grains cm^{-2} yr^{-1}) from the beginning of organic sedimentation, at roughly 13,500, and although the shape of the curve of birch influx plotted against time is similar to that for the other sites referred to above, the chronology is roughly one millennium older. It has always been puzzling that the pollen sequences at the Tuktoyaktuk sites begin directly with high birch values and that there is no evidence of an initial herb pollen assemblage that might have represented a pioneer vegetation on the immediately deglaciated landscape. There is no obvious solution to this problem. As mentioned in Chapter 2, the rolling, irregular terrain of the Tuktoyaktuk Peninsula south and southwest of the settlement of Tuktoyaktuk has given the geologists difficulty in ascertaining its origin. The latest consensus appears to be (Geological Survey of Canada, Map 32, 1979) that it formed by extensive late Pleistocene thermokarst modification of an older morainic landscape. The lakes, including those at T-5 and Sleet, are relatively deep with moderately steep but stable shores, quite unlike more recent, active thermokarst lakes. But it is possible that they began their existence in the late Pleistocene as thermokarst collapse events which obscured any early pollen stratigraphy and caused slight discrepancies in radiocarbon age. Against such an argument one should point out that the ages for the oldest sediments from these sites fit well the chronology of deglaciation developed independently by Fyles et al. (1972) and Rampton and Bouchard (1975). The problem awaits further investigation.

We can sum up the birch evidence by stating that, with the exception of the Tuktoyaktuk Peninsula data, it appears to support hypothesis 2a: a rapid, major climatic warming at 11,000 yr BP.

2.5. HEATH AND MIRE VEGETATION

Ericad pollen is underrepresented in most lake sediments and therefore its frequency is often too low and discontinuous to make useful interpretation possible. However, Cwynar (1982) deliberately both counted large enough pollen sums (>1,000 per sample) and made taxonomic separations within the Ericales so that he was able to decipher important vegetation changes. His herbs-and-subshrubs diagram (his Figure 6, reproduced here in simplified form as Figure 30) together with his influx diagram (reproduced here in simplified form, Figure 21) illustrates clearly that, in parallel with the large increases in dwarf birch influx, there was a marked increase at about 11,000 in heath representation and influx, and he reconstructs the vegetation at the period 11,000–8,900 as follows: 'the increase of these taxa (*Ledum, Cassiope, Vaccinium, Empetrum*) must reflect the local development of wet-mesic heath communities, an interpretation supported

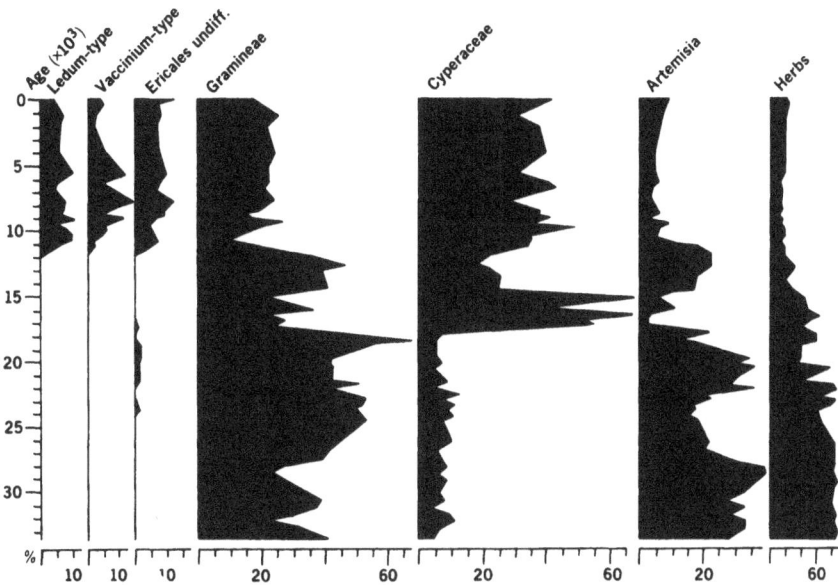

Figure 30 A summary percentage pollen diagram for Hanging Lake, based on a pollen sum excluding tree and shrub taxa, after Cwynar (1982, Figure 6)

by the corresponding increase of *Sphagnum* spores and the occurrence of *Rubus chamaemorus*, a plant of wet heaths.'

He then points out that the coeval 'seven-fold increased influx of Cyperaceae probably indicates the development of widespread tussock tundra on the surrounding plains.' He proposes greater summer precipitation and increased summer temperatures to explain these and other changes at about 11,000 yr.

Following Cwynar's lead, a similar effort to count large numbers per sample of the Lateral Pond site samples proved to be useful in revealing the development of a heath-tundra element at about 11,000 yr (Ritchie 1982).

A plot of the natural logarithm of five-sample running means of the pollen influx of the main pollen taxa at Lateral Pond (Figure 31) illustrates several of the points made above. At about 11,000 there were rapid changes: a decline in the herb and grass curves and a rapid increase in dwarf birch followed by increases in spruce and ericads.

This evidence for a marked shift from herb tundras to dwarf birch tundras and mires at about 11,000 supports the hypothesis of a climatic control involving increased summer warmth and increased precipitation. It is also possible that slow soil development by humus accumulation and leaching of nutrients would have promoted the development of heath tundras, but the rapid rate of the change suggests climate rather than soil as the predominant factor.

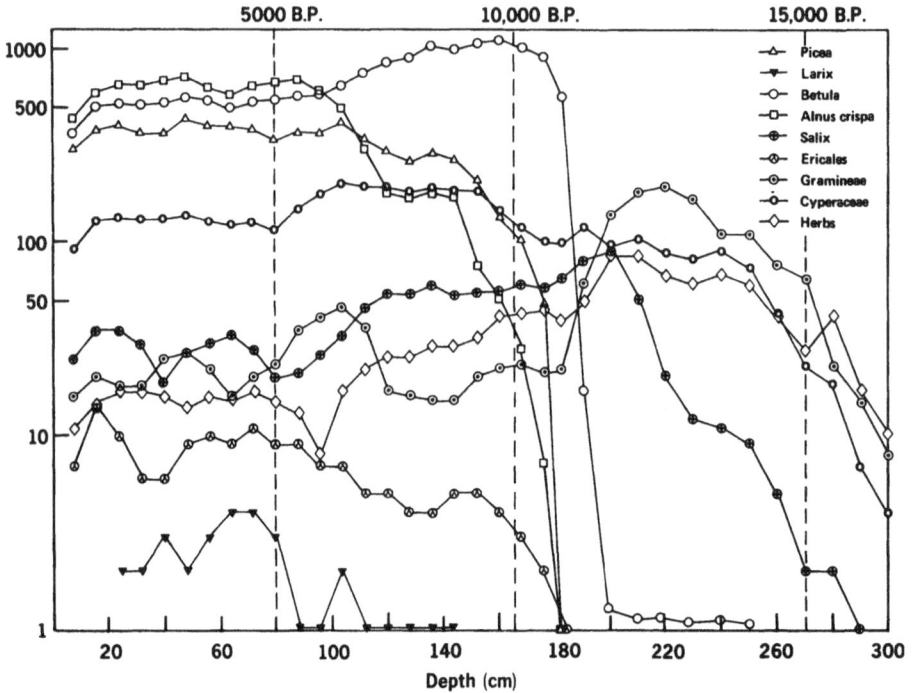

Figure 31 The natural logarithm of five-sample running mean pollen influx values of the main taxa plotted against depth and approximate radiocarbon age for the Lateral Pond site. It shows clearly that between 12,000 and 10,000 rapid and major changes in influx occurred, with more gradual changes before and after that transitional period between the latest glacial and the present interglacial. The vegetation reconstruction proposed is from a treeless landscape with sparse herb tundra on uplands and sedge-grass marshes and willow scrub in the lowlands in the late-glacial to the modern landscape with spruce forests on lower and upper slopes, tundra on ridges, and mires in the lowlands.

2.6. TYPHA

This pollen type is rarely frequent enough in sediments to produce continuous percentage curves, but its occurrence as individual grains is of possible palaeoenvironmental value because it has been shown by detailed field measurements that its average pollen dispersal distance is small, in the range 2–10 m, with maximum values of 1 km in strong winds (Krattinger 1975). It follows that the occurrence of *Typha* pollen in primary lake or mire sediments can be taken confidently as an indication of presence of the plant in the immediate vicinity. Its modern distribution suggests that it has climate indicator value (Porsild and Cody 1980, Figure 48). It is a plant of almost unlimited, cosmopolitan range, the exception being that it has a northern limit in North America. In northwest Canada the limit lies

Table 13 Time range in radiocarbon years of the occurrences of *Typha* pollen at the 12 sites shown in Figure 29. Note that each line indicates only the time range and does not signify continuous records throughout the particular period.

Site, number on Figure 29 and reference	Radiocarbon yr BP × 10⁻³								
	12	11	10	9	8	7	6	5	4
1 Hanging Lake (Cwynar 1982)	▬▬								
2 M-Lake (Ritchie 1977)		▬▬							
3 TT-Lake (this volume)			▬▬▬	▬					
4 KE, Old Crow (Ovenden 1981)			▬						
5 BL, Old Crow (Ovenden 1981)			▬						
6 SW-Lake (this volume)			▬▬▬	▬▬▬	▬▬▬	▬▬			
7 Sleet Lake (Spear 1983)		▬▬▬	▬▬▬	▬▬▬	▬▬▬	▬▬▬	▬▬▬	▬▬	
8 Mayday Lake (Spear unpubl.)		▬	▬						
9 Hendrickson Is. pingo (Hyvärinen and Ritchie 1975)				▬					
10 Eskimo Lake pingo (Hyvärinen and Ritchie 1975)			▬▬▬	▬					
11 Richards Island pingo (Ritchie 1972)			▬▬	▬▬▬	▬▬▬	▬▬▬	▬▬		
12 East Channel site (Zoltai and Ritchie unpubl.)			▬▬						

roughly 75 km south of the southern boundary of our study area. It is common in the Mackenzie Valley as far north as Norman Wells and it extends westwards to a northern limit in Alaska near Fairbanks. Thus, if we assume that its range has always been in equilibrium with climate, its occurrence in postglacial sediments enables us to use it as an indicator of past climate. In making this assumption we should note carefully that range alterations in time might have been controlled also by changes in hydric environments that occurred independently of climate.

Pollen occurs in late Pleistocene and Holocene sediments at 12 sites widely scattered in the area (Figure 29); the time range of *Typha* for each site is shown in Table 13. Clearly it grew commonly in the area between 10,000 and 9,000 yr.BP, but also occurred as early as 12,000 and persisted at three sites until 5,000. An example of a continuous *Typha* pollen record was found in six samples of marly peat collected by S. Zoltai from the base of an eroding high-centred polygon mire at 69°05′N, 134°08′W, 8 km east of the East Channel of the Mackenzie River (Figure 32). The unit has a radiocarbon age greater than 7,510, the age of fossil *Larix* wood found immediately above the samples shown in Figure 32 (Zoltai and Zalasky 1979).

If we assume that climate is the sole determinant of distribution, a conservative estimate of the climate in the region to the immediate east of the Mackenzie Delta for the period 11,000–9,000 is a summer regime warmer by about 6°C for the mean of the months May, June, July, and August, and by about 350 degree-days above 5°C for the year, than it is at present. These figures represent the differences between the modern values for Norman Wells and Dawson on the one hand and

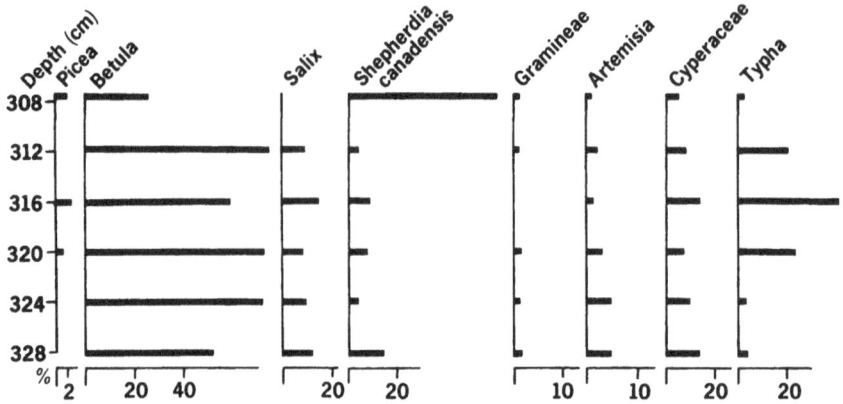

Figure 32 A percentage pollen diagram based on analyses of samples from an exposure of peat in a high-centred polygon at 69°05′ N, 134°08′ W (site 12, Figure 11). A sample of larch wood from above this unit, at the 298 cm level, gave a radiocarbon age of 7,510 (Zoltai and Zalasky 1979). The pollen sum included only terrestrial types so the *Typha* percentages were calculated outside the sum.

Inuvik on the other (Table 1). In other words, the *Typha* evidence provides strong support for hypothesis 2a, that the climate at 12,000–10,000 was significantly warmer than at present.

2.7. MYRICA

Similarly, but at fewer sites, the frequencies of *Myrica* pollen are supportive, as Cwynar (1982), Ovenden (1981), and Spear (1983) have suggested. *Myrica* is represented in the northwest by a single species, *M. gale*, a circumpolar boreal species. Its modern northern range limit is the East Channel of the Mackenzie River, and it is absent from the modern Tuktoyaktuk Peninsula and North Yukon. If we assume that its northern limit is climatically determined, and that it is and has been throughout the Holocene in equilibrium with climate, then the *Myrica* pollen curves at sites beyond its modern range are indicative of climate change. Sleet Lake provides clear evidence there (Figures 13, 14): its percentage and influx peaks are confined to the interval 9,000–6,500 with a maximum influx value of 200 cm^{-2} yr^{-1} at 9,000. The time range at Tuktoyaktuk-5 is similar but the amounts are less. Similar *Myrica* maxima and ranges occur at other sites: at Hanging Lake, between 10,500 and 7,000; at TT-Lake a peak at 8,000 and a range of 8,000–6,000; at SW-Lake, a range from 8,500 to 4,500; and at M-Lake a minor peak at 8,500, though sporadic grains occur from then to the youngest samples.

The *Myrica* data can be explained by the following minimal climatic changes: that during the period from 9,000 to 6,000 the climate at the coast of the Tuktoyaktuk Peninsula was as warm as the modern Inuvik climate, and that the modern conditions were reached following a cooling that began at about 8,000 and terminated at about 5,000.

2.8. COMMUNITY DYNAMICS

Several sites in the forest zone show rapid changes in the pollen influx of willow, poplar, juniper, and spruce during the interval 10,500–8,000, immediately after what we have concluded above was a climatically controlled surge in pollen influx at 11,500. The influx changes can be seen readily, for example, in the diagrams from M-Lake, TT-Lake, and SW-Lake (Figures 33, 34, and 35).

Hypothesis 2e proposes that these relatively rapid changes can be explained independently of a climatic cause. In particular, I suggest that they were controlled by the population ecology of the main taxa, and that the herb–dwarf birch tundras that occupied most upland sites at 11,000 were invaded by ecologically localized populations of balsam and aspen poplar, tall shrubby willow species, and juniper, to form a mosaic of vegetation types on upper mesic slopes, with extensive balsam poplar groves on shoreline and alluvial substrata; aspen groves and juniper stands probably occupied the driest and sunniest upland surfaces. The arrival of spruce and the development of increasing areas of closed conifer forest constricted the areas of juniper, aspen, and willows because of their intolerance of increased shade. The timing of this ecological event was controlled by the arrival of migrating spruce populations, and we have already derived the mean radiocarbon age of that as 9,100.

The alternative explanation, hypothesis 2c, is that climate, or some related environmental variable, changed during the period 11,000–8,000. For example, a slight cooling might have caused a decrease of less cold-tolerant aspen poplar and juniper, particularly *Juniperus horizontalis*, in favour of the hardier *Picea glauca*. A related explanation might be that slight cooling after 11,000 decreased the depth of the soil active layer and caused changes in the proportions of the dominants. Alternatively, as suggested above, these vegetational changes might have been caused by changes in the seasonal distribution of warmth and / or precipitation.

These alternative explanations can be evaluated by examining in detail the influx changes of the main taxa for the period in question. Graphs of the influx of spruce, poplar, birch (dwarf), willow, and juniper against radiocarbon age for TT-Lake, SW-Lake, and M-Lake produce a pattern of change that is remarkably similar at all three sites (Figures 33, 34, and 35). Dwarf birch influx rises from about 100 at 11,000 yr to roughly 2,000 at 10,500–10,000 yr, then drops moderately to between 500 and 1,000 at 8,500 yr. Poplar, willow, and juniper values increase sharply from <10 to ~100 between 10,500 and 9,500 yr, but juniper lags behind willow and poplar by about 500 yr at TT and M lakes. At about 9,000 yr these three taxa show sharp declines, coincident with a rapid rise in spruce influx from <10 at 9,500 yr to ~500 at 8,500 yr.

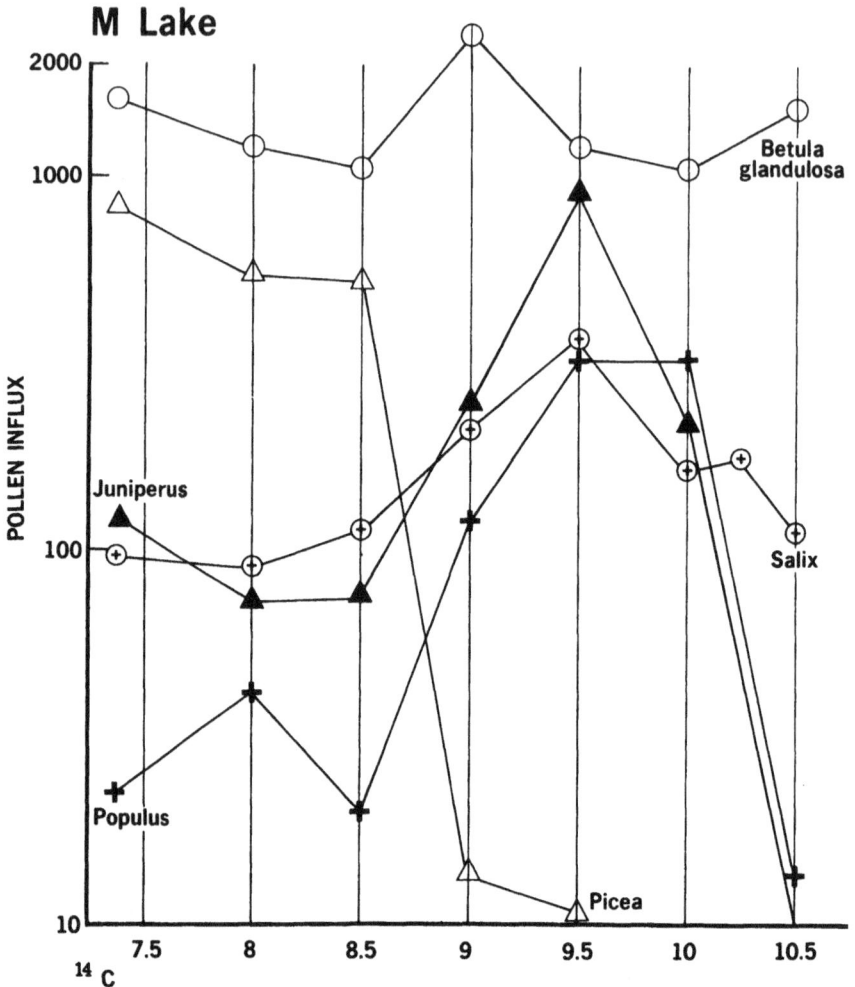

Figure 33 The natural logarithm of pollen influx of the taxa shown plotted against interpolated radiocarbon age

Hypothesis 2e explains the changes between 9,500 and 8,500 yr as the result of the competitive disadvantage of shade-intolerant species of poplar, willow, and juniper in a spruce-forested landscape. It might be added that the record of *Shepherdia canadensis* pollen (see, for example, Figure 19) could be explained in a similar way although, being an entomophilous species, its pollen is underrepresented and often does not yield large enough continuous frequencies to produce useful graphs.

In this discussion I am assuming that the Cupressineae pollen type, indistinguish-

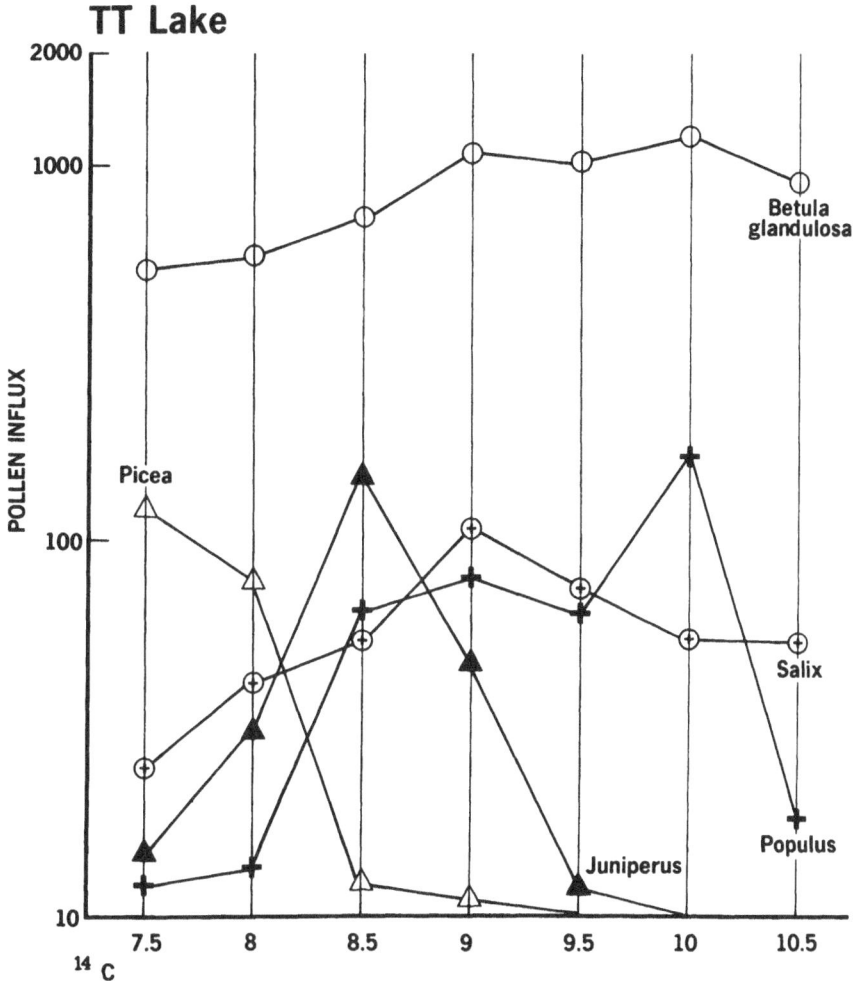

Figure 34 The natural logarithm of pollen influx of the taxa shown, plotted against interpolated radiocarbon age

able on morphological grounds, can be assigned certainly to *Juniperus* rather than *Thuja* because the latter is totally absent from the Northwest Territories, the Yukon, and Alaska. Further, I assume that all the *Juniperus* records are of *J. communis* rather than *J. horizontalis. J. communis* grows in the area today, and it is found on open, unshaded, well-drained slopes in the boreal forest region but only rarely beyond the spruce treeline. It occurs abundantly on sunny, south-facing slopes where the white spruce canopy is discontinuous because of topographic irregularity (Plate 22), and in such communities it is characteristically associated with *Potentilla fruticosa, Shepherdia canadensis, Arctostaphylos uva-ursi,* and

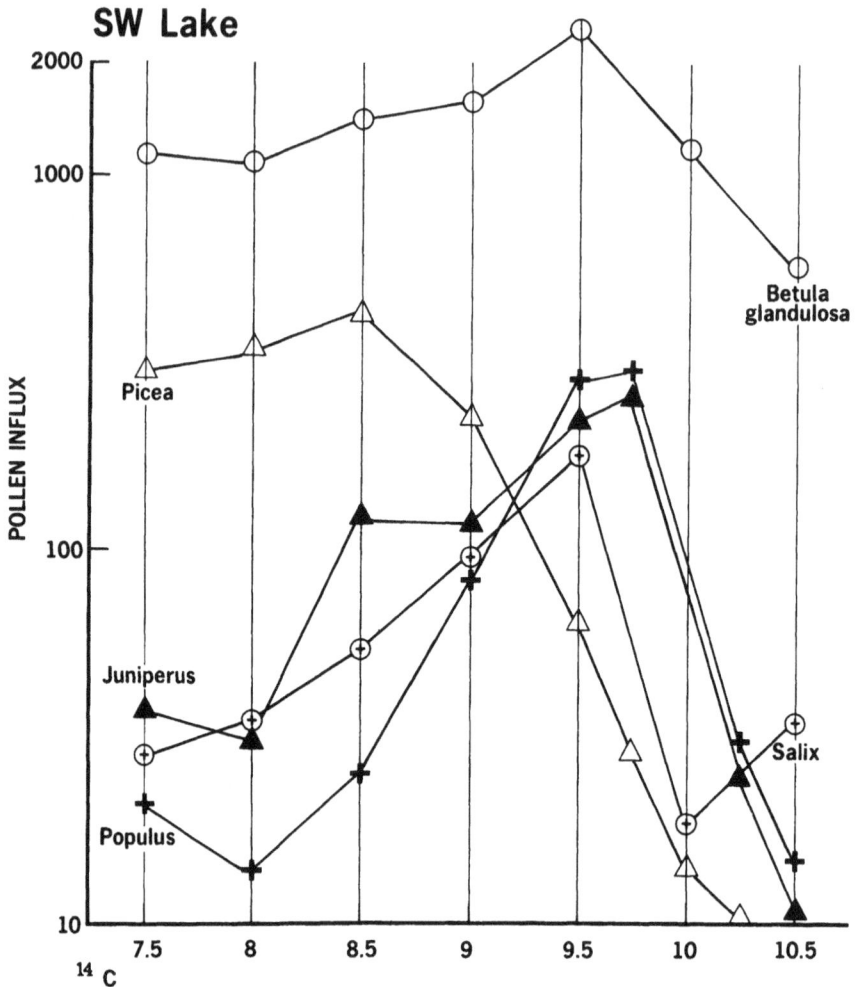

Figure 35 The natural logarithm of pollen influx of the taxa shown, plotted against radiocarbon age

Dryas octopetala. Occasionally it grows with small clonal trees of *Populus tremuloides* on exposed, south-facing knolls where the spruce canopy is discontinuous. In areas where calcareous bedrock prevails, for example in parts of the South Richardson Mountains and the Keele Range, and especially where fire has not occurred frequently, the landscape is mantled with continuous, pure spruce forest, and only exposed scarps and steep sunny slopes where the canopy is discontinuous (Plates 10 and 22) have a different cover – usually scrubby aspen, juniper, and tall willow.

It is also probable that the configuration of the poplar and spruce can be explained in part as a reflection of a large-scale successional change on the vast alluvial soils of the nearby Mackenzie Delta. In fact the graphs of basal area against time for balsam poplar and spruce plotted by Van Cleve et al. (1980, Figure 9) for the Tanana River successional sequence are remarkably similar to the pollen curves in Figures 33–35. It is likely that the arrival of white spruce in the delta at 9,100 precipitated extensive replacement of balsam poplar, by the same process that occurs locally in the modern delta on higher alluvial surfaces (Gill 1973c).

I must stress that these are observational, anecdotal comments on the ecology of these taxa, and they await critical field analysis and experiment for confirmation or otherwise. However, if accurate, they suggest that a replacement by spruce due to its superior competition for light is adequate to explain the pollen data. A danger in speculation of this sort is that we adjust the *scale* of the pollen signal to suit our particular explanation. Here I am suggesting that the pollen changes reflect changes at the level of plant community, or at most segments of the landscape (land-form–vegetation units). In fact we have no secure basis in modern pollen vegetation data to permit such interpretive subtleties, although the problem is familiar and will no doubt be subject to continuing research (Webb 1981; Prentice 1983; Delcourt et al. 1983). The fact that we find similar responses at widely spaced sites in the same region (TT-Lake, M-Lake, SW-Lake, K-Pond) suggests that a more general process, possibly mediated by climate change, controlled the shifts in the mosaic of land-form–vegetation patterns.

Hypothesis 2c would gain support if we assumed that the *Juniperus* was *J. horizontalis*, or that both species were present in the late Pleistocene. The northern limit of *J. horizontalis* barely extends into the study area, and it appears that it is climatically restricted. A surge and then decline of *J. horizontalis* pollen between 10,000 and 8,500 would be concordant with the hypothesis of a cooling at 9,000–8,500. However, *J. horizontalis* is restricted to calcareous substrata, at least in its northwestern distribution, and it is unlikely that the SW-Lake juniper curve could be ascribed to *J. horizontalis* since the rock types and derived soils of that region are non-calcareous. *J. communis*, in contrast, occurs on both calcareous and non-calcareous soils, and its chief apparent habitat characteristics are open, unshaded surfaces, often south-facing, with freely drained, usually coarse-textured soils.

While hypothesis 2e (climatic change) cannot be refuted by the available data, it is difficult to find useful autecological data to test it adequately. The modern ranges of many shrubby boreal willows, the two poplar species, and *Juniperus communis* overlap with that of *Picea glauca* and therefore one assumes that their macroclimatic limits are similar. However, the persistence of poplar, juniper, willows (and *Shepherdia canadensis*) in open habitats within the modern boreal forest is compatible with hypothesis 2d.

A similar phenomenon is observed in the influx curves from the Lateral Pond site (Figure 31), but there the participation of poplar and juniper is not significant while non-arboreal elements are more important, presumably because Lateral Pond is in a mountainous region with extensive extant tundras while the three sites considered above are all in the Mackenzie Valley region, within the modern boreal forest.

We will return in Chapter 7 to these major vegetation changes that occurred in the interval 11,000–8,000 and consider them in a broader ecological context.

3. The Late Holocene Alder Rise and Subsequent Stability

The final biostratigraphic events requiring explanation are the abrupt rise in alder pollen percentages recorded in all Holocene diagrams from the area and the subsequent absence of significant changes in frequency of the main pollen taxa. As the percentage diagrams in Chapter 5 show, the alder percentage rises from >1% to maxima between 30% and 65%, and this change delimits the lower boundary of the latest pollen zone, occurring in our area about 6,700 radiocarbon years ago (range 5,700–7,800), at least one to two millennia after the spruce pollen rise. In addition, most diagrams show an interval with low alder frequencies (~5%) prior to the main alder rise.

Two alternative hypotheses to explain the alder data have been proposed by Hopkins et al. (1981) and I offer a third; they are numbered here 3a, 3b, and 3c.

Hypothesis 3a suggests that alder 'became extinct in Beringia as early as 45,000 or 50,000 yr ago' (Hopkins et al. 1981, p 240) and reinvaded in the early Holocene from a refugium in coastal British Columbia or Washington. I suggest a supplementary part to this explanation, that it might also have survived south of the Laurentide ice and then migrated to the northwest along the Cordilleran foothills and Mackenzie Valley route during the early Holocene.

Hypothesis 3b proposes survival in the northwest throughout the latest glacial cycle, in scattered sites, 'persisting mainly as parthenogenetic shrubs in small isolated thickets ... ,' following which 'the climatic conditions permitting release and proliferation of widespread alder thickets were, according to the pollen data, delayed until about 7,000 yr ago' (Hopkins et al. 1981, p 240).

Hypothesis 3c is an extension of 3a above in that it proposes that the alder taxa were absent from the entire northwest during full- and late-glacial times, that they migrated into the area in small numbers at roughly the same time as spruce, and that the abrupt increase of alder pollen at about 6,700 yr (in our area) was a response, mainly by *A. crispa*, to an opening up of the closed spruce forest as its northern margin responded both to the slow climate cooling that followed the thermal maximum at 9,000 and to the cumulative effect of natural fires.

Before we evaluate the alternative explanations, let us assemble some basic facts that might be relevant.

Three alder taxa occur in northwest North America. *Alnus crispa* is found throughout boreal and subarctic North America, extending into Greenland and East Asia, and it reaches its maximum abundance in the open spruce forests and forest-tundra landscapes that form the broad ecotone between the shrub tundras of the low-arctic and the closed spruce forests of the boreal zone. In northwestern

North America this species extends slightly beyond both the arctic and montane treelines. For example in North Yukon it is common on the rolling uplands northeast of the Old Crow Flats, beyond the limit of continuous open spruce woodlands. On the Tuktoyaktuk Peninsula it extends 20–40 km north of the limit of spruce treeline, occurring as scattered shrubs on upland surfaces in dwarf birch heath tundra communities, chiefly on lower slopes with a deeper snow accumulation rather than on upper slopes and summits. In the Richardson Mountains, scattered *A. crispa* form a narrow zone extending 10–20 m upslope from the upper limit of spruce occurrence. In all these communities at the limit of its range it flowers and sets seed annually, as it does throughout its range (Gilbert and Payette 1982; Patterson et al. 1983). A recent investigation of the ecology of *A. crispa* at the treeline in Quebec by Gilbert and Payette (1982) led them to hypothesize that certain populations expanded locally beyond their previous northern limit in response to a warm period, between 1920 and 1960, but the critical factor, they suggest, was the presence of open, mineral soils resulting from periglacial phenomena. They conclude that in the absence of an open mineral substratum with minimal competition from other plants, alder will spread actively only vegetatively although seed formation takes place normally.

A. crispa ssp. *sinuata* is an amphiberingian race, morphologically poorly distinguished from *A. crispa*, confined to submontane upland habitats in South Alaska and adjacent coastal British Columbia, as well as Kamchatka and East Siberia. It is absent from our area, and the nearest localities are at 64°N in western Yukon.

A. incana ssp. *tenuifolia* is a North American boreal-subarctic species confined to western Canada and adjacent Alaska. It does not extend beyond the limit of closed boreal forest, and in our area its northern limits are the Mackenzie Delta in the east and the Porcupine River in the west. It is confined to alluvial habitats.

A. crispa and *A. incana* pollen types have been identified by some workers recently (Cwynar 1982; Ritchie 1982; Brubaker et al. 1983; Ovenden 1981) using the criteria developed by Richard (1970).

Maximum percentages of alder pollen in modern samples occur in the forest-tundra ecotone (Table 14). In our area they comprise over 40% of the pollen sum, while in central Canada where pine is common, with its pollen making up 20–60% of the pollen sum, the average alder frequency in the forest tundra is 17%. We have already noted in Chapter 5 that alder is the most overrepresented of all pollen types in this area, with pollen percentage: cover percentage ratios ranging from 10:1 to 20:1. Therefore, assuming no other major changes in the pollen spectrum, the alder pollen rise from values of 1–5% to maxima at 30–65% *implies a vegetational change of minor proportions* – from a landscape with very rare alder bushes (or none if we assume the low pollen frequencies are all of long-distant origin) to one with a maximum total alder cover of not more than 5%.

Let us consider first hypotheses 3a and 3b. The sub-fossil pollen record from our area (Table 15, Figure 36) is similar to the spruce data in that there is no

Table 14 Alnus pollen percentages in modern lake sediments from north-central and northwestern Canada based on Lichti-Federovich and Ritchie (1968) and Ritchie (1974)

	x	SD	*n*	*x*	SD	*n*	*x*	SD	*n*	*x*	SD	*n*
Mackenzie sites	31.4	6.7	15	44.8	12.0	8	not represented			28.0	2.3	9
Keewatin, N. Manitoba, N. Saskatchewan sites	11.9	7.9	14	16.8	9.5	10	15.6	5.6	14	8.1	4.6	52
		TUNDRA			FOREST-TUNDRA			SPRUCE-PINE FOREST			SPRUCE FOREST	

Table 15 Radiocarbon age of the 10% alder threshold at 17 sites in northwest Canada and Alaska. The values in parentheses indicate the beginning of *Alnus incana* values greater than 2%

Site and reference	Latitude	Longitude	Age (yr BP)
Eildun Lake, Slater (1980)	63°08'	122°46'	7,500
Sweet Little Lake, Figure 19	67°38'	132°00'	5,700
Lateral Pond, Ritchie (1982)	65°57'	135°31'	6,800 (8,000)
Tyrrell Lake, Ritchie (1982)	66°03'	135°39'	7,500
M-Lake, Ritchie (1977)	68°16'	133°28'	6,600
Twin Tamarack Lake, Figure 16	68°18'	133°25'	6,000 (7,200)
Tuktoyaktuk-5, Ritchie and Hare (1971)	69°03'	133°27'	6,000
Sleet Lake, Spear (1983)	69°17'	133°35'	7,000
Eskimo Lake pingo, Hyvärinen and Ritchie (1975)	69°25'	131°40'	6,800
Hendrickson Island, pingo, Hyvärinen and Ritchie (1975)	69°32'	133°35'	6,800
Hanging Lake, Cwynar (1982)	68°23'	138°23'	7,800 (9,000)
Polybog, Old Crow, Ovenden (1981)	67°48'	139°48'	6,500 (7,200)
Birch Lake, core I, Ager (1975)	64°19'	146°40'	7,700
Birch Lake, core II, Ager (1975)	64°19'	146°40'	8,200
Lake George, Ager (1975)	63°47'	144°30'	7,800
Alatna Valley, Brubaker et al. (1983)	67°30'	153°30'	7,000
Antifreeze Pond, Rampton (1971)	62°21'	140°50'	5,600

obvious directional trend in the radiocarbon ages of the threshold value. I have chosen 10% arbitrarily as the value, and the data show as much variation between adjacent sites as between distant sites. At the four sites where the *A. crispa* and *A. incana* pollen types have been distinguished, small frequencies (<5% of the pollen sum) of *A. incana* were registered roughly 1,000 yr before *A. crispa* appears, suggesting that the speckled alder preceded the upland species into the northwest.

The most detailed macrofossil record (Spear 1983) confirms the presence of *A. crispa* by 6,500 yr on the Tuktoyaktuk Peninsula, and shows continuous presence from then to the present in that area. The Twin Lake, Inuvik, data (Fyles et al. 1972) show alder leaves and cones at depths with a radiocarbon age greater than

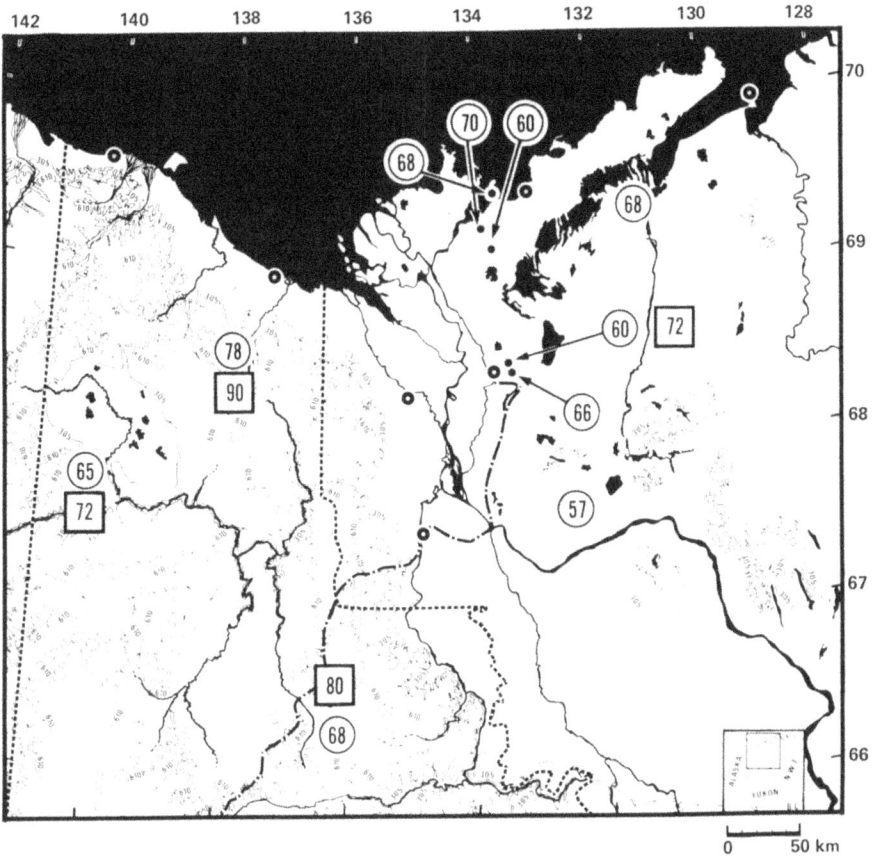

Figure 36 The circled figures are radiocarbon ages in hundreds of years for the sites shown, of the 10 per cent threshold alder pollen. Values in open squares refer to the earliest continuous curve of *Alnus incana*, regardless of percentage. Site locations are listed in Table 15.

8,200 yr, almost 2,000 yr before the alder pollen rise in the same section and in adjacent sites. This apparent discrepancy led Matthews (1975b) to propose that alder had been in the area for one or two millennia but had produced no pollen, persisting solely by vegetative propagation. This of course did not explain the presence of seeds at Twin Lake so parthenogenesis was evoked in a later explanation (Hopkins et al. 1981). Parthenogenesis has not been recorded in northern species of *Alnus* and, as I have noted above, modern treeline specimens flower and set seed abundantly. It is possible that this single anomalous alder record is unreliable and should be discounted, since all other alder records in this region show no age discrepancy between the pollen and seed data. However, if hypothesis 3c is supportable it is possible that isolated plants of alder reached the delta area of the lower Mackenzie River at roughly the same time as spruce, in which case the Twin Lake site record is acceptable, though still unusual.

Alder is the most prolific pollen producer in the northwest so it is unlikely that it could have survived during several millennia prior to ~7,000, including the full-glacial, without leaving any sub-fossil record. However, a Pacific coast refugium for *A. crispa* ssp. *sinuata* seems both geologically feasible and concordant with the pollen record summarized by Hopkins et al. (1981), although this does not account for the *A. crispa* and *A. incana* data from our area. All the pollen influx and other evidence we have reviewed above points to a significant climatic warming at ~11,000. It is quite unlikely that sexual reproduction in surviving plants of *A. crispa* or *A. incana* would have been delayed for four or five millennia, until about 6,500 BP. Therefore, I conclude that hypothesis 3b is untenable in the light of present, admittedly inconclusive data.

Hypotheses 3a and 3c call for migration from a southern refugium. The data are inconclusive. The Lofty Lake record (Lichti-Federovich 1972) has the alder rise at 8,000 and that would imply a migration to the lower Mackenzie area of over 1,800 km in roughly 1,400 yr. However, the John Klondike Lake site, just north of the Alberta boundary at 60°21′N, 123°38′W (Matthews 1980), gives 8,700 as the date of the alder rise, in which case a migration rate of 1,000 km in 2,000 yr is inferred. The latter seems reasonable, although no quantitative data on seed dispersal in alder exist. It is likely that the riverine species *A. incana* spread downstream relatively rapidly and might well have reached the delta in small numbers by 8,000 yr, followed later by *A. crispa*. Hypothesis 3a seems reasonable though poorly documented, but it does not account for the lag between the alder rise and the spruce rise since the taxa have very similar macroclimatic requirements and seed dispersal characteristics.

Hypothesis 3c accommodates the lag by proposing that small numbers of alder plants, *A. incana* on the alluvial sites and *A. crispa* on uplands, reached the study area from a southern refugium at roughly the same time as spruce. It then requires that this invading spruce forest, predominantly white spruce, spread quickly in a warm climate – the warmest of the present interglacial – to its northern limit by 9,000, as a closed canopy forest, limited in the east of our area by the Tuktoyaktuk Peninsula coastline and the wet lowlands of the distal part of the peninsula, and in the west by the British and Barn mountains. *A. crispa* occurred in this forest, but infrequently, as it does in modern closed-canopy forests, yielding the low pollen frequencies recorded in the diagrams prior to the abrupt alder rise. Two factors then took effect: a slow cooling of climate, and the occurrence and cumulative effect of natural fires. They caused an opening of the northern parts of the spruce forest to produce the first occurrence of a forest-tundra zone, or a forest-tundra ecotone. This opening up of the spruce canopy stimulated alder to spread into unshaded habitats, increasing its total representation from <1% to about 5% of the upland vegetation, an adequate increase to produce the large increase in alder pollen percentage, to roughly the same as its modern cover in the forest-tundra

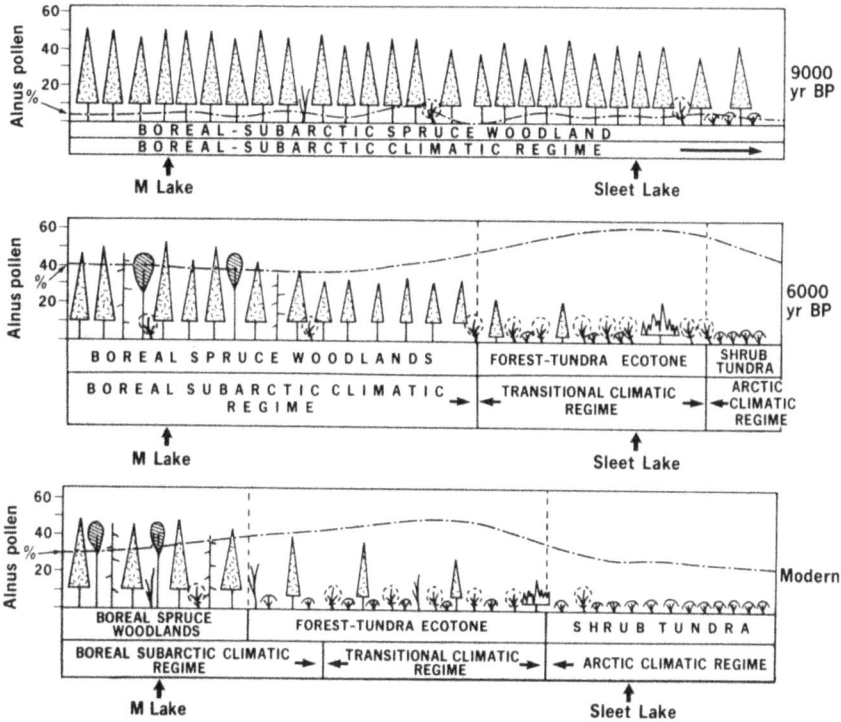

Figure 37 A highly schematic, hypothetical profile diagram through the modern, 6,000, and 9,000 zonation along a transect from the vicinity of M-Lake to beyond Sleet Lake, to account for the alder rise at 6,000 in terms of an opening of the northern margin of the spruce woodlands

east of the Mackenzie Delta (Corns 1974; Reid 1974). A schematic caricature of this entirely speculative explanation is attempted in Figure 37.

Indirect support for this idea that there has been a reciprocal relationship between the amount of alder in the landscape and the density of spruce cover is found in a recent investigation by Richard (1981) of the Holocene vegetation history of Ungava. There the chronology and sequence of the main pollen taxa are quite different from those of western North America, but he shows that migration of a dense spruce-tundra vegetation into areas that had been occupied by alder–dwarf birch tundra caused a diminution of alder pollen frequencies.

Hypothesis 3c is highly speculative, with several weak points. It depends on the assumption that the white spruce communities that advanced north rapidly at ~9,000 BP formed closed-canopy stands on most upland sites. The high spruce influx values recorded at the northern sites support this notion, and the modern ecology of white spruce (Zasada and Gregory 1969; Viereck 1970; and others) indicates that it is a rapid colonizer of mesic, mineral soils, forming closed stands unless some moisture deficiency, as on coarse-grained sandy soils, or soil temperature effect, such as a high permafrost table, prevails. However, there is no firm palaeobotanical evidence for or against this assumption about the stand structure of these white spruce communities. Hypothesis 3c also depends on the assumption that the ecological amplitude of *A. crispa* is slightly greater than that of spruce, in that it grows today slightly farther north in the arctic and slightly higher in the mountains. The explanation (Figure 37) suggests that as the northern fringe of spruce woodlands experienced less favourable Holocene conditions, moderately after about 8,000 and then more abruptly at about 4,500, regeneration of spruce became suppressed, areas of shrub tundra expanded at the expense of trees, and spruce persisted as reduced, often krummholz plants reproducing vegetatively. These generally cooling, but fluctuating, climatic conditions resulted in a fringe of discontinuous spruce woodlands, and the cumulative effect of natural fires further suppressed arboreal growth in favour of shrubby forms, including alder. Black and Bliss (1978) and Viereck and Schandelmeier (1980) have reviewed the evidence for such changes, and Elliott (1979) has assembled the palynological and dendrochronological data that suggest that the western arctic treeline is today, and was at times during the Holocene, out of equilibrium with climate. In contrast, Garfinkel and Brubaker (1980) report that, in the Central Brooks Range of Alaska, white spruce 'is not greatly stressed at its present limit and that seedlings are abundant above treeline.'

Green (1981, 1982) has demonstrated the type of response implied in hypothesis 3c, but for different species in a different geographical region. He shows close correlations between pollen influx and charcoal influx values that suggest that a recurring major perturbation – in this case large fires with a frequency of 1 in 300 yr – had the effect of opening up the vegetation to permit the expansion of species that had been present in very low amounts. A similar phenomenon might be visualized for the alder rise phenomenon in our area, at least as a hypothesis for future investigation. It suggests that large fires in the fuel-rich closed spruce forests that occupied the Tuktoyaktuk Peninsula between 9,000 and 7,000 yr opened up the forest and permitted a rapid expansion of alder. The slow deterioration in climate that set in shortly thereafter, and then the more rapid cooling at 4,500, retarded the restoration of a closed forest. Fires recurred, as they do today, but on a smaller scale because of the reduced fuel accumulation of the forest-tundra transition zone, further promoting the patchiness that is one of the characteristics of this zone.

In conclusion, as Hopkins et al. (1981) pointed out, the alder record is both too sparse and too ambiguous to encourage a confident advocacy of any of the three hypotheses discussed above. Moreover, it is likely that additional explanations will emerge as new palaeobotanical data accrue. For example, if the early palynological intimations from the Sleet Lake (Spear 1983), Alatna Valley (Brubaker et al. 1983), and SW-Lake (this volume, Chapter 5) sites, that the early Holocene spruce forests were predominantly white spruce communities and that

black spruce became more important later, prove to represent a general phenomenon, then hypotheses are likely that suggest widespread edaphic changes,
possibly stimulated by cooler and moister climates, promoting paludification and
rising permafrost tables. Such reconstructions might well also accommodate the
alder pollen data, one of the many stimulating avenues of investigation that lie
ahead, as Gilbert and Payette (1982) emphasize in their recent study.

Before leaving the alder record we should note that Davis (1981) calls attention
to the same phenomenon in a large set of pollen diagrams from eastern North
America. She writes (p 140), 'everywhere spruce preceded alder, the pollen of
which peaked after spruce began to decline.' She makes no further comment, but
it is clear that a sequence of events as visualized in hypothesis 3c would not
explain the eastern North American data, where many more tree species are
involved in the Holocene vegetation history, in a different climatic regime. The
alder problem awaits solution.

The second aspect of the late Holocene record of interest is the apparent
stability of the vegetation at all sites during the latest five millennia. Even the
sensitive Sleet Lake site, situated very close to the modern treeline and with close
interval samples and carefully estimated pollen concentration and influx values
(Spear 1983), reveals no significant fluctuations during the past at least 4,000 yr.
These results are of interest because they are not in agreement with data from
treeline sites to both the west and east of our area. For example, Brubaker et al.
(1983) report that '*Picea glauca* apparently did not reach its modern limit in the
Alatna Valley [Central Brooks Range] until 1,500 years ago.' In central Canada,
Sorenson (1977) interprets fossil podsol evidence from sites in Keewatin in terms
of a forest border at 3,500 yr BP 280 km north of its modern position. In total, he
reconstructs tentatively the treeline history of southwest Keewatin in terms of six
major displacements: 280 km north at 4,500 yr, ~50 km south at 3,000 yr, ~50 km
north at 2,500 yr, ~50 km south at ~2,000 yr, 75 km north at 1,200 yr, and ~60 km
south at 900 yr.

Gagnon (1982) uses a large array of radiocarbon-dated fossils of larch wood to
trace treeline movements in an area of Nouveau Québec between western Ungava
Bay and Richmond Gulf. As we noted above, larch is a sensitive indicator of
climate change because of its dependence on seed reproduction. He records a
maximum expansion northward of larch treeline between 3,500 and 2,700 yr, a
lesser advance at 2,000–1,600, and regressions at 2,700–2,400, 2,100–2,000, and
1,600–1,300. These are almost exactly the opposite changes to those reported by
Sorenson (1977) for the middle of the continent for the spruce treeline.

The pollen evidence from central Canada is inconclusive. Nichols (1976)
interprets data from peat bogs in terms of several changes in treeline during the
past four millennia, while Kay (1979), based on analysis of pollen in lake sediments, finds a major advance north of the present position between 5,500 and
3,700 yr BP and little change since that episode.

Even if adequate data were available, this would not be the appropriate place to review the Holocene history of the North American arctic treeline. But a few points should be made as pertinent commentary on the apparent lack of evidence from our area of immediate study.

1. Modern pollen records from sites across the boundary between forest and tundra show that in this area of steep climatic gradients (see Chapter 2) and a narrow forest-tundra transition, small changes in treeline position might not be registered significantly in the pollen record because of the large regional component in the pollen rain (Ritchie 1974).

2. As we have seen in Chapter 2, there is a correlation between the forest-tundra boundary and the frequencies in summer of dry continental arctic and warm humid Pacific masses, a point that was first made by Bryson (1966). Hare (1968) has pointed out that the steep gradient in the summer thermal conditions across the northwest treeline (see Figure 2) is due to the relatively narrow width of the frontal zone in this part of the continent, by contrast with the central regions to the immediate west and east of Hudson Bay. The Yukon-Alaska mountains, and such dynamic features of the atmospheric circulation as the Aleutian low, serve to anchor or focus the air-mass wave patterns as they traverse the northwest of Canada. In other words, smaller amplitudes of Holocene latitudinal variation of treeline are to be expected in our region than, for example, in southern Keewatin.

3. The detailed and careful investigations of Payette and his co-workers in subarctic Quebec (e.g. Payette and Gagnon 1979) provide several interesting and cautionary points. The most striking is the demonstration that black spruce can maintain its treeline position in spite of deteriorating climate by active vegetative growth, while larch responds directly and immediately to change because of its total dependence on seed reproduction. A second point they make, in the form of a hypothesis, is that frequent natural fires within the forest border during the late Holocene have opened up the formerly closed forests to create a wider forest-tundra zone, independently of climate change.

Clearly, more critical investigations are required in our region before a coherent reconstruction of the latest millennia will be possible. We will return to this topic of future research needs in the next chapter.

To summarize this chapter, I offer the following reconstruction of the palaeoenvironments of the northwest. No doubt some readers will have recognized additional phenomena of interest in the palaeobotanical data of Chapter 5, and it is certain that other explanations than those I have singled out can be devised to explain the various phenomena. It is to stimulate such responses that I have deliberately kept quite separate the presentation of the basic facts and my attempts to draw inferences.

1. During the glacial period between 115,000 and prior to the final stadial event at ~25,000, the climate went through at least one cycle of conditions from colder than present-day to almost as warm and back to colder, roughly comparable to the differences between what are described in Chapter 2 as arctic coastal and

subarctic continental climates. However the biostratigraphic record for this time-span remains discontinuous, equivocal, and inadequately dated.

2. During the latest glacial cycle, between 24,000 and 11,000, the climate of the ice-free uplands and valley systems was much colder and drier than at present, being generally similar to modern mid-arctic conditions. Modern values for central Banks and Victoria islands provide an approximation of the regional palaeoclimate, but I stress that much micro- and meso-climatic variation existed, as we will discuss in the next chapter, and that the distinctive circulation patterns of full-glacial times (Lamb and Woodroffe 1970) would have produced conditions with no exact modern analogue.

3. The regional climate, particularly in the growing season, warmed slowly between 15,000 and 12,000, after which there was a rapid warming to conditions warmer than at present by not later than 10,000. Mean July temperatures were 3–5°C higher than modern values, with proportional increases in the length of the growing season, degree-day totals, and the length of the frost-free period. It is possible that episodes of seasonally distinctive climate occurred within the early Holocene, reflected in the pollen data from sites in the forest zone.

4. The period of warmer-than-modern conditions lasted until ~4,500, but a slow cooling began at ~8,000.

5. The climate cooled more rapidly at 4,500 when the modern climatic regime was established, and there have been only minor fluctuations since that time.

Some changes in the palaeobotanical record occurred independently of these palaeoclimatic variations, particularly those due to differences in the life history characteristics and ecological responses of several of the dominant species.

It will be noted (number 3 above) that I support the interpretation of McCulloch and Hopkins (1966) that proposes that the maximum summer warmth of the present interglacial was reached by 9,000 or 10,000 yr ago, and that this appears to be at variance with the northwest European data (e.g. Lamb 1977 gives 8,000–5,000) and a recent summary of the Northern Hemisphere evidence (Budyko 1982, 'about 6,000 yr ago'). However, as we noted earlier, we can expect significant differences in both the chronology and the sequence of climatic changes in different segments of the Northern Hemisphere, due to differences in the amplitude and position of the wave components of atmospheric circulation. Further, we should expect significant differences between northwest Canada and eastern North America for the additional reason that between 10,000 and 8,000 yr BP, when the northwest was completely free of both the direct and indirect effects of Laurentide and Cordilleran ice, eastern North America lay immediately to the south and east of a massive waning continental ice sheet that covered roughly 3.5 million km^2 and was probably responsible for the delay of several millennia in vegetational response to the general warming. Indeed, we should expect northwest North America to have minimal vegetational response delays and to show the closest agreement with such data as the 18-O curves, which indicate a temperature maximum at 9,000 (Stuiver 1970).

The Milankovitch theory of climate forcing provides for a maximum of summer radiation in high latitudes at ~10,000 yr BP (Berger 1981). It appears that the particular area of our study is ideally situated to provide sensitive registrations of these expected effects of climatic change, by contrast with areas to both the west and east. Sites in Alaska immediately south of the Brooks Range (Brubaker et al. 1983) do not show a clear spruce pollen maximum in the early Holocene, possibly because of the complex effects of large orographic features. The large network of sites in eastern North America that provide records for the entire Holocene appear to reinforce the traditional conclusion that the 'thermal maximum' occurred in the mid-Holocene; the reason for this delayed and possibly reduced effect is probably that when the maximum warmth occurred the central and eastern parts of Canada were covered by a thick mass of Laurentide ice (Prest 1970) with resulting cooling effects on the westerly air flows. By contrast northwest Canada, being upwind of the ice mass, would have experienced the maximum effect of the summer radiation maximum.

The palaeobotanical data assembled above provide the first convincing corroboration from North America of the early Holocene thermal maximum.

7
Current Problems and Future Trends

1. Ecological Implications of the Palaeobotanical Record

When a botanist addresses palaeontological and archaeological questions there is a risk of entering intellectual minefields, just as palaeontologists, geologists, and archaeologists can and do devise false constructs when they dabble with pollen diagrams. So I will try to avoid the temptations of scientific dilettantism and confine myself to a consideration of the following questions that depend primarily for their resolution on the palaeobotanical record:

1. Is the reconstructed full-glacial and late-glacial vegetation of northwest Canada concordant, ecologically, with the faunal record?

2. Does the palaeobotanical evidence provide useful, testable hypotheses to explain major changes in the faunal, particularly vertebrate, record?

Pleistocene mammal bones have been collected in northwest Canada for over 100 years. The vast number of specimens (several tens of thousands), large number of species (about 60), and extinct status of many of the animals have generated both intense scientific interest and a romantic, popular aura, the latter analogous to the lure of gold-mining, which, as it happens, led to the unearthing of most of the fossils. The scientific thrust was described by Canada's authority on Beringian palaeontology as follows: 'A concerted effort to study the abundant and interesting ice-age vertebrate fossils of the Yukon region began in 1966, when long-term projects near Dawson and the Old Crow Basin were initiated by the National Museum of Canada' (Harington 1978, p 52). At the same time, placer gold-mining in the Fairbanks area of interior Alaska, and later a series of scientific studies, yielded a similar rich harvest of fossil bones.

The first serious attempt to consider the palaeoecological significance of these enormous collections of mammalian fossils was also the most influential in establishing a widespread conception of northwest Quaternary environments: a perceptive, germinal analysis by Guthrie (1968). His central conclusion, quoted in full below, has had a profound influence on the course of Beringian palaeoecology:

This study of the palaeoecology of four fossil assemblages of large mammals from the late Pleistocene sediments near Fairbanks, Alaska, emphasizes the structure, composition, habitat, and the pattern of subsequent extinction of the community. All four faunas were composed predominantly of grazers. Bison, horse, and mammoth were the most common species. Many component species of this complex community of large mammals became extinct near the close of the Wisconsin glaciation, leaving the comparatively depauperate community that exists in Alaska today. The high percentage of grazers in the fossil community suggests that Interior Alaska was a grassland environment during the late Pleistocene.

From our immediate, botanical point of view, the gist of Guthrie's argument was threefold: (a) that many of these assemblages of fossils represented *contemporaneous* animal communities, (b) that the vast majority of the animals were grazers, and (c) that these communities existed during glacial times, in particular the latest glacial cycle. Geologists (Hopkins 1972, 1979; Matthews 1976), other zoologists (Hoffmann 1974; Harington 1978), archaeologists (Morlan 1980; Irving and Harington 1973), plant ecologists (Bliss 1975; Young 1976), and others embraced the Guthrie hypothesis enthusiastically and it quickly became entrenched in both popular (Canby 1979) and serious compilations. As happens so often when interesting and provocative ideas are injected into the scientific milieu, few readers seem to have considered carefully the original caveats. The first was emphasized as the leading sentence of the main section of his paper (p 347), stating: 'Unfortunately, the extensive series of Alaskan fossil large mammals were collected with no attempt to relate the specimens with the physical stratigraphy.' Later a further warning was sounded (p 350): 'Whether or not all species were exact contemporaries, at least they all inhabited the area during part of, or throughout, a very short geological time span.' Then the crucial question of animal density – 'one of the more difficult problems posed by this assemblage is the matter of density' (p 354) – was circumvented by an oblique linking of apparent high diversity to numbers of animals, with the only quantitative substantiation a table of percentages of different ungulates in the four fossil assemblages, each an amalgam of provenances derived from sediments that were 'mainly Wisconsin age silt deposits, but pre-Wisconsin sediments present' (Table I and comment on p 350). His conclusion (p 355) was: 'Rather than being comparable to the densities found on the present tundra, the ungulate production in the Alaskan refugium was perhaps several times as great.'

And the final conclusion: 'Thus one might conclude that the climate change accompanying the last glaciation, along with other factors, resulted in a widespread grassland environment in the northern Holarctic region and supported a rather extensive and complex large-mammal "grassland" community.'

Thus was born, on these tenuous data and large assumptions, the Pleistocene grassland biome, or steppe-tundra concept, and, as is the case with all imaginative ideas, it stimulated much fruitful discussion and new research on a problem that still awaits complete understanding.

Schweger and Habgood (1976) sounded a warning that the assumption of contemporaneity of vertebrate fossils was unfounded, and subsequent research in the Old Crow Basin and elsewhere has failed to provide the essential biostratigraphic evidence that different species were in fact living in community (Morlan 1980). Only the Bluefish Cave deposit (Cinq-Mars 1979) provided secure evidence that mammoth, horse, bison, dall sheep, muskox, and elk were coeval in North Yukon, but the maximum radiocarbon age of this deposit is 15,500 yr.

It transpires that the full-glacial vertebrate record is particularly sparse, and certainly not adequate to support the concept of a 'mammoth-steppe' biome. The only radiocarbon-dated bones of full-glacial age from northwest Canada are a specimen of *Bison crassocornis* from the Dawson area (I-3570, age 22,200 ± 1,400) and one of *Mammuthus primigenius* from the Old Crow River area (I-3573, age 22,600 ± 600), both reported by Harington (1978). All other fossils of these mammals, and of horse, reindeer, dall sheep, and elk, are either of younger or older age. Other ecologically interesting taxa such as camel, scimitar cat, saiga antelope, American lion, and extinct muskox (*Symbos*) are either of unknown Pleistocene age or 'pre-Wisconsin' age, although Harington (1980) assumes in the absence of dates or secure stratigraphic data that six saiga fossils from northeast North America, including one from Baillie Island at 70°35′N, 128°06′W, are all of Wisconsin age.

So two central parts of the Guthrie hypothesis (contemporaneity and full-glacial age) rest on slim or otherwise unsatisfactory evidence.

The third point, that a grassland vegetation must be reconstructed for full-glacial times to reconcile the assumed food preferences of the dominant mammals, can be tested against three data sets: the dentition and other morphological traits of the animals that suggest certain feeding behaviour; the botanical content associated with the fossils that might give a direct indication of diet and vegetation; and the palaeobotanical record of coeval sediments. I am not competent to assess the first line of evidence and I refer the reader to Guthrie (1982) for an entry point to the pertinent literature.

The only directly associated botanical evidence comes from Eurasian mammoth fossil sites, and analyses of stomach content (Tikhomirov 1958; Solonovich et al. 1977) and of the associated sediments (Solonovich et al. 1977, Shilo 1978; Shilo et al. 1977; Vereschagin 1977; Kubiak 1982) have revealed a wide diversity of taxa (*Betula, Larix, Salix*, Ericaceae, Gramineae, Cyperaceae, and many herb genera) and a large age range (12,000–180,000). These results suggest that woolly mammoth was not an obligate grazer, and the evidence cannot be used reliably to indicate a treeless, grassland environment. On the contrary, most of the Eurasian mammoth fossils are associated with botanical evidence that suggests an environment little different from modern subarctic or arctic situations in the northwest.

The final data set, the palaeobotanical record, has expanded rapidly in recent years as more sites in Alaska and northwest Canada have been studied in detail. We will examine three data sets to test the grassland hypothesis against the palaeobotanical record.

1.1. THE OLD CROW BASIN SEDIMENTS

The pollen assemblage zone from the Pleistocene riverbank sediments that shows the lowest arboreal and highest grass proportion, and is therefore *prima facie* the one most likely to provide evidence to support a grassland hypothesis, is the pollen assemblage III, and to a lesser extent II, compiled by Lichti-Federovich (1973) from the analyses we referred to in some detail in Chapter 5. In particular her site 2 is of interest because it is the same as locality 44 of Harington (Irving and Harington 1973; Harington 1978) and it has produced a wealth of mammal bones, including mammoth, horse, camel, and caribou, but also some small mammals of modern arctic and boreal affinity (Table 16). Harington (1978) suggests that this unit is of Sangamon age because most of the radiocarbon age determinations gave infinite values, and because 'analysis of plant macrofossils and invertebrates indicates that the climate was as warm as at present.' Matthews (1975a, 1982) has summarized his largely unpublished study of insect and plant macrofossils from these and similar deposits, and while he consolidates the botanical data into ecologically uninformative groupings such as the tantalizing '*Potentilla*-Cruciferae complex,' it seems that he has found no firm botanical evidence of strictly grassland elements. Lichti-Federovich's (1973) type III assemblage remains the only botanical evidence from these sites with some suggestion of a grassland element, particularly the high percentages of grass pollen. However, if we compare the pollen spectra from these sites with unequivocal Holocene grassland assemblages from a pair of sites in southern Manitoba (see Table 16, abstracted from Ritchie and Lichti-Federovich 1968), it is clear that there are important, conclusive differences. The Holocene spectra provide a more reliable comparison than those from modern grassland sites because the latter include a significant proportion of tree and shrub pollen derived from extensive prairie tree plantations and the natural grassland element is greatly suppressed by the prevalence of cultivated land. In addition to the obvious differences between these spectra as listed in Table 16, it is important to note that the Holocene grassland spectra contained pollen of true grassland indicator species (*Sarcobatus, Sphaeralcea, Phlox hoodii, Collomia, Amorpha canescens, Symphoricarpus albus*) while the Pleistocene sediments from the Old Crow Basin include such boreal and arctic indicator elements as *Shepherdia canadensis, Rubus chamaemorus, Saxifraga, Polemonium,* and *Selaginella.* These distinctions led Lichti-Federovich (1973) to conclude that the assemblages represented 'a treeless landscape with extensive tundras, dominated by sedges and grasses, with substantial herb components' for type III, and 'a forest-tundra type ... with spruce on

Table 16 A summary of the mammal fossil records and the pollen assemblages recorded from unit 2, locality 44, at 68°12' N, 140°00' W on the Old Crow River. The mammal data were abstracted from Harington (1978; pp 62–3) and the pollen data were summarized from Figure 3 in Lichti-Federovich (1973). Average pollen frequencies for the grassland assemblage zone in Holocene lake sediment from two sites in southern Manitoba are included, abstracted from Ritchie and Lichti-Federovich (1968).

Mammal taxa	Pollen taxa	Pollen percentages (sum = all terrestrial taxa)		Holocene grassland	
		Assemblage Type II	Assemblage Type III	1	2*
Ochotona princeps (pika)	Picea	10–45	10–15	1–2	1
Lepus arcticus (arctic hare)	Betula	15–40	5–20	1–2	1
Spermophilus parryi (arctic ground squirrel)	Alnus	2–15	1–5	2–5	2–3
Castor canadensis (beaver)	Salix	2–10	1–5	1–2	1–2
Castoroides ohioensis (giant beaver)	Juniperus	trace	2	trace	1–2
Dicrostonyx torquatus (collared lemming)	Ericaceae	2–4	1	none	none
Lemmus sibiricus (brown lemming)	Cyperaceae	15–30	10–20	10–15	2–5
Clethrionomys cf. rutilus (red-backed vole)	Gramineae	10–20	40–60	20–30	25–30
Ondatra zibethicus (muskrat)	Chenopodiineae	trace	1–2	10–20	10–15
Microtus xanthognathus (chestnut-cheeked vole)	Artemisia	trace	1–2	20–30	25–40
Alopex lagopus (fox)	Ambrosieae	none	none	10–20	5–10
Mustela sp. (weasel)					
Gulo gulo (wolverine)					
Spilogale sp. (spotted skunk)					
Mammuthus sp. (mammoth)					
Equus cf. (Plesippus) verae (large horse)					
Rangifer tarandus (caribou)					

*1 = average values from pollen zone III at Belmont site; 2 = average values from pollen zone III at Glenboro site. Pollen zone III at both sites has a radiocarbon age range of 4,000–9,000.

alluvial and other similar sites with deep active layers, while interfluves were occupied by a rich shrub-herb tundra; willows and local alder stands occurred along watercourses, and local heaths and dwarf birch tundra occupied stable, humous substrata' for type II.

Curiously, these perceptive interpretations, based on detailed analyses, have been largely ignored by subsequent students of the 'mammoth-steppe-tundra biome' question, although no data have accrued from the subsequent intensive investigations of Old Crow, Bluefish, or Bell Basin sediments that have significantly advanced or modified them.

The central difficulty has been that the riverbank sections consist of predominantly redeposited sediments, so not only is it impossible to establish contemporaneity of different mammal species, but it is rarely clear that a particular pollen-bearing unit was coeval with a certain fauna.

However, let us assume purely in the interests of exploring the hypothesis that most of the Old Crow locality 44 fossils represent a set of coeval animals, and let us further assume that the pollen zone II and III assemblages (Table 16) were contemporaneous with these faunal elements. Are the data sets concordant? A conclusive answer, of course, is impossible as nothing is known of the ecology of the extinct animals, but it might be reasonable to suggest that the diverse range of vegetation types available in a boreal or subarctic type climate could have been productive and varied enough to sustain a varied vertebrate fauna. Of course that is ecologically a rather bland statement, and we will pursue it more rigorously later in this chapter, but it suggests that some of the Pleistocene ecosystems represented by the northwest fossil record could have existed in *non-glacial* episodes, when a varied, productive vegetation of both arboreal and non-arboreal elements prevailed. The corollary is that the preoccupation of some palaeoecologists with the search for a coherent association of animals and plants to occupy a *glacial*-episode landscape has been focused too narrowly. It is instructive to note that the vast majority of comparable Eurasian fossil assemblages show an association of the same or similar fauna and a diverse palaeobotanical record indicating both treeless and treed communities (Kubiak 1982; Müller-Beck 1982).

1.2. THE FULL-GLACIAL POLLEN RECORD

The concept of a productive grassland vegetation in full- and late-glacial times was questioned in light of the late-glacial pollen influx data from M-Lake (Chapter 5), and I sounded some reservations by suggesting that the reconstruction based on herb-zone pollen spectra resembled arctic polar desert *physiognomically* and that it probably represented a herb-dominated tundra on upland soils (Ritchie 1977). This view was amplified considerably in light of the detailed full-glacial and late-glacial records from Hanging Lake and then the late-glacial record from Lateral Pond, both discussed in Chapter 6 and in full in the original publications.

We concluded (Cwynar and Ritchie 1980) that at least in eastern Beringia the herb zone 'represents a sparse, discontinuous vegetation of herbaceous tundra on upland sites and local sedge-grass meadows on lowlands.' Our prediction that 'further work in Alaska will show that the herb zone has very low influx values' appears to be supported by data from the Brooks Range (Brubaker et al. 1983) and elsewhere (P. Anderson, personal communication). We suggested that in addition to and as important as the influx record, the long list of pollen taxa recorded in these sediments that have clear arctic-alpine modern ranges is further support for the sparse herb-tundra hypothesis. So far no full-glacial pollen or plant macrofossil evidence has been published from North Yukon sites that supports an alternative hypothesis to our sparse herb-tundra reconstruction. The main alternative is a grassland vegetation, as originally proposed by Guthrie (1968) and adopted by several other authors, but, as we have seen above (Table 16), none of the diagnostic ecological characteristics of a grassland has been recorded from any Pleistocene sediments examined in eastern Beringia.

However, these alternative hypotheses may not be as disparate as they seem. Part of the difference is semantic. If the concept grassland is being used in the original sense of Guthrie (1968), implying a continuous, productive grassland landscape similar to modern prairies of the Western Interior of North America, then the botanical evidence from full- and late-glacial sediments and from the older Old Crow and Bluefish Basin sections does not provide support for such a reconstruction. However, if the term steppe-tundra is being used to describe the vegetation component of this biome in the sense used by Russian botanists, then the differences of interpretation are more apparent than real. As we shall see below, steppe-tundra as used in the Russian literature describes a plant community that North American ecologists describe as herb tundra, and it consists of a community of predominantly herbaceous plants including grasses and sedges. It is difficult to ascertain exactly what vegetation is proposed by the proponents of the steppe-tundra concept, particularly as it is assumed by them that the biome is now extinct and therefore has no modern equivalent.

Perhaps the most productive steps are to continue the search for modern analogues for the full- and late-glacial pollen spectra; to persist in the investigation of Pleistocene sediments in the hope that fossil assemblages will be found in secure, datable stratigraphic position; and, as we propose in the next section, to attempt to interpret those few instances where there is unequivocal evidence of contemporaneous mammal and herb-tundra fossils. The search for a modern analogue does not imply, as Guthrie (1982, p 309) asserts, an adherence to the notion of 'cliseral shifts of biotic provinces ... through time.' On the contrary, as Prentice (1983) shows in a recent review, plant palaeoecologists have long realized that many full- and late-glacial assemblages lack modern analogues and that major taxa often behaved individualistically throughout parts of the record. At least one investigation from eastern Beringia (Ritchie 1977) supports that viewpoint.

1.3. BLUEFISH CAVES

Cinq-Mars (1979) has reported on preliminary analyses of the palaeontological and archaeological evidence, and Ritchie et al. (1982) on the pollen, found in loess and humus deposits in two limestone caves in the Bluefish River valley of North Yukon. I have dealt with the pollen record already, in Chapter 5. This is the only site in northwest Canada with secure evidence linking a late Pleistocene fauna of mammoth, horse, dall sheep, caribou, bison, muskox, and elk (wapiti) with pollen evidence for a treeless vegetation. The radiocarbon age of this assemblage is 15,500 ±130 (GSC-3053) to 12,900 ±120 (GSC-2881). The pollen data agree well with those from other late-glacial sites in North Yukon, and the vegetation reconstruction suggested is therefore probably as reliable as can be derived from cave sediments. It indicates a landscape covered by a sparse herb tundra on upland surfaces and a complex of sedge-grass marshes with willows on lowlands. These data do not support the original Guthrie hypothesis, unless we use the concept grassland rather loosely. However, they are not necessarily discordant with the concept 'steppe-tundra' as used by our Russian colleagues, so let us digress briefly to clarify this point, raised already in Chapter 4 in discussing tundra vegetation types.

Russian botanists use the term true steppe to describe grass- and herb-dominated zonal communities that characterize continental climatic zones with warm summers, and they are homologous with northern prairies or grasslands of the Western Interior of North America. The term cryophytic steppe (synonymous with tundra-steppe and arctic-steppe) is used to describe azonal communities dominated by grasses and herbs that occur on well-drained, often south-facing slopes in the Siberian arctic and subarctic. Full descriptions of these terms and concepts can be found in Lavrenko and Sochava (1956), Aleksandrova (1980), and Yurtsev (1972, 1974, 1978, 1981). An example from our area will illustrate the differences between Russian and North American nomenclature. Local, well-drained, sometimes south-facing knolls and escarpments, with coarse-textured soils, occur throughout our area, and they support a varied non-arboreal, quite localized community in which the following species are common members: *Pulsatilla multifida, Calamagrostis purpurascens, Carex supina* ssp. *spaniocarpa, Plantago canescens, Artemisia frigida, Erigeron compositus, Braya humilis.* For example, bedrock escarpments along the north shore of Dolomite Lake near Inuvik have several stands of this community (Plate 10); local examples can be found throughout the southern portion of the Tuktoyaktuk Peninsula on the gravel soils of kame knobs (Corns 1974); and they are common on south-facing riverbanks and terraces throughout North Yukon (Cwynar 1982). Russian botanists would refer to these as examples of 'relict cryophyte-steppes' (or tundra-steppes, or arctic steppes), and detailed descriptions, illustrations, and maps can be found in a recent monograph by Yurtsev (1981). North American ecologists, however, would describe them as local herb-tundra facies, or by some similar designation, but the terms steppe and relict would not be used, (a) because although *physiognomically* they resemble grasslands, *none of the species is geographically or ecologically a grassland or prairie type*; on the contrary, all have continuous arctic and / or subarctic ranges in North America; and (b) because there is no evidence, palaeobotanical or distributional, that any of these species is in fact of relict status.

It follows that any palaeoenvironmental reconstruction of a herb-tundra assemblage should not invoke a steppe concept that is equivalent to grassland in North American usage, unless there is strong supporting evidence of grassland elements in the palaeobotanical record. It would be helpful, though it seems a vain hope, if Freitag's (1977) recommendation, that the term steppe should be used only in its strict, original sense, were adopted, but it is encouraging that one Beringian palaeoecologist has decided to abandon the term (Colinvaux 1980).

So when we searched for an appropriate reconstruction of the Bluefish Cave pollen record, and of all late-glacial pollen records from our area, the floristic composition and low pollen concentration led us (Chapter 5) to propose a herb-tundra community complex, and it was suggested that the closest modern equivalents were the communities on South Banks and South Victoria islands in the western Canadian arctic. I further suggested (Chapter 6) that the modern climates of these areas might provide an approximate analogue for the late-glacial environments of the North Yukon, though it was stressed that it is unlikely any modern analogue will match closely late-glacial times with their unique juxtaposition of montane glaciers, extensive sea ice, lowered sea level and exposed continental shelves, and distinctive synoptic-scale circulation patterns.

However, let us, again in the interests of exploratory discussion, assume that the late-glacial (15,500–12,500 yr BP) climate and vegetation of the Bluefish River region were similar to what is found today on South Banks and South Victoria islands. *Would such a reconstruction be concordant with the palaeontological record?*

Two tests of the proposition are possible.

First, does the composition, diversity, and density of the modern fauna of these areas provide a valid basis for pursuing the idea? Two large mammals are common on Banks and Victoria islands today, caribou and muskoxen, but we should note that their numbers have changed markedly in historical times: decreases in caribou, and the opposite in muskox. These populations have not been studied as intensively as those farther north, on the Queen Elizabeth Islands (Miller et al. 1977; Parker and Ross 1976; Parker 1978; Thomas et al. 1981; and others). However, early surveys by Manning (1960) and Macpherson (1961) and a recent survey by Jakimchuk and Carruthers (1980) provide basic information on density and distribution in relation to vegetation, while the Queen Elizabeth investigations give essential data on dietary preferences and limiting environmental factors. Jakimchuk and Carruthers (1980) estimate that, at the time of their survey (1980), there were ~12,000 muskoxen and ~8,000 caribou on Victoria

Island, and that the main concentrations were on the west side, in the regions of Prince Albert Sound and western Prince Albert Peninsula. Caribou, mainly the Peary caribou race, show a preference for sites of intermediate elevation, in response to their foraging preferences: in summer, *Salix arctica*, *Dryas*, and other herbs; in winter, mosses, willow, and sedge-grass. Muskox, in contrast, prefer lowland sites, particularly in summer, when sedge-grass meadows are their most important habitats and *Carex*, *Eriophorum*, and herbs are the chief forage species. Topographic diversity seems to be important in determining the winter distribution of muskoxen, related to shelter, and to browse and graze availability as a function of snow cover. The different but overlapping forage preferences of the Peary caribou and muskox are assumed to account for their non-competitive sharing of range in the arctic (Parker 1978), at least during times when conditions are not severe.

Banks Island supports the largest number of muskoxen (18,000), roughly one-third of the entire Canadian population, but there is a smaller Peary caribou number (<1,000). It is of interest to note the observation of Jakimchuk and Carruthers (1980) that large areas of suitable caribou habitat in southern and eastern Victoria Island are not utilized at present, and it is likely that it represents a habitat 'vacuum,' left after the extirpation by Inuit hunters, in the early decades of this century, of the Dolphin and Union herds which migrated annually between the mainland and Victoria Island, and whose numbers were estimated at between 20,000 and 100,000 animals (Manning 1960).

It is difficult to estimate accurately average density values for these areas. Miller et al. (1977) give 1.8 km^{-2} for muskoxen in the Thomsen River valley of Banks Island, and conclude that in general 'the best muskox habitat below 200 m in Canada seemingly can support 1–2 muskoxen per km^2 on a year round basis if the winters are not too severe.' They document severe population declines in muskoxen correlated with winters of severe conditions, particularly with deeper than average snowfall and thicker than average ice crusts on the tundra. Caribou densities are difficult to estimate for these areas but 1–2 km^{-2} is probably a conservative value. In addition to these large mammals, the terrestrial fauna includes large populations of arctic fox and lemmings, and smaller populations of arctic hare and wolf.

I conclude, quite tentatively, that the modern putative analogue for the late-glacial of North Yukon does support a relatively large, but not particularly diverse, fauna of herbivores.

The second test of the hypothesis that the late-glacial landscapes surrounding Bluefish Caves could have supported a fauna to match the fossil records is to develop the general hypothetical carrying-capacity estimate made by Bliss and Richards (1982) for full-glacial Beringia into a specific estimate for the Bluefish region. Bliss and Richards (1982) have applied the most accurate available data on primary and secondary production to derive estimates of the potential numbers of six herbivore species supportable by a prescribed array of hypothetical full-glacial vegetation. The steps in the calculations are simple: net annual production of vegetation → annual harvest by herbivores → consumption of plant biomass by each species from each vegetation type → estimates of animal biomass and numbers, assuming certain rates of reproduction, mortality, and harvesting.

In the case examined here the past vegetation of the 1,000 km² surrounding the Bluefish Caves site was mapped on the basis of the pollen data from samples spanning 12,500–15,500, making certain assumptions about the relationships between topographic position and vegetation type, following the reconstructions developed earlier for this site (Chapter 5, and Ritchie et al. 1982) and in more detail for a very similar region (Lateral Pond Site, Chapter 7 and Ritchie 1982). The reconstruction (Figures 38 and 39) assumes that lowland areas adjacent to streams and lakes at elevations between 300 and 400 masl were occupied by sedge-grass marshes and willow thickets; that upland ridges above 600 m were occupied by discontinuous fellfield and cushion plant tundras; and that intermediate slopes between 400 and 600 m were covered by a herb-tundra sward, similar to the modern *Dryas–Carex scirpoidea* tundra communities described under (b) in Chapter 4, Section 2.2.

I follow Bliss and Richards (1982) and assume that the lowland habitats consist of one-third willow cover and two-thirds sedge-grass, and I adopt directly their vegetation type utilization estimates for the six mammal species that occur at Bluefish, which entails replacing moose with what I assume for present purposes is the closely comparable wapiti. I have omitted dall sheep from the calculations since little information is available on which to base estimates. A schematic caricature of the Bluefish area landscapes and their utilization is attempted in Figure 39.

Similarly, following their review of net above-ground annual primary production data for high-latitude vegetation types, I have selected lower values than they used for their full-glacial Beringia exercise, partly to give a minimal, conservative estimate, but also because the values chosen are close to modern values for mid-arctic sites, namely 200 g m^{-2} for lowland willow shrub, 50 g m^{-2} for sedge-grass meadow, 10 g m^{-2} for herb tundra, and 5 g m^{-2} for fellfield and cushion tundra (Webber 1974). The equivalent values in Bliss and Richards's reconstruction were 400, 75, 100, and 25 respectively.

The calculation of carrying-capacity estimates is set out in Table 17; it follows four steps as follows, corresponding to the numbered sections of the table:

1. The area estimates and above-ground primary production values were derived as described above; their product yields the net available production.

2. The percentage of above-ground production actually utilized for each community type (15%, 15%, 10%, and 5%) follows Bliss and Richards's (1982) assessment of values in the literature. We should note carefully that these values might be high, and in any event they are based on large assumptions.

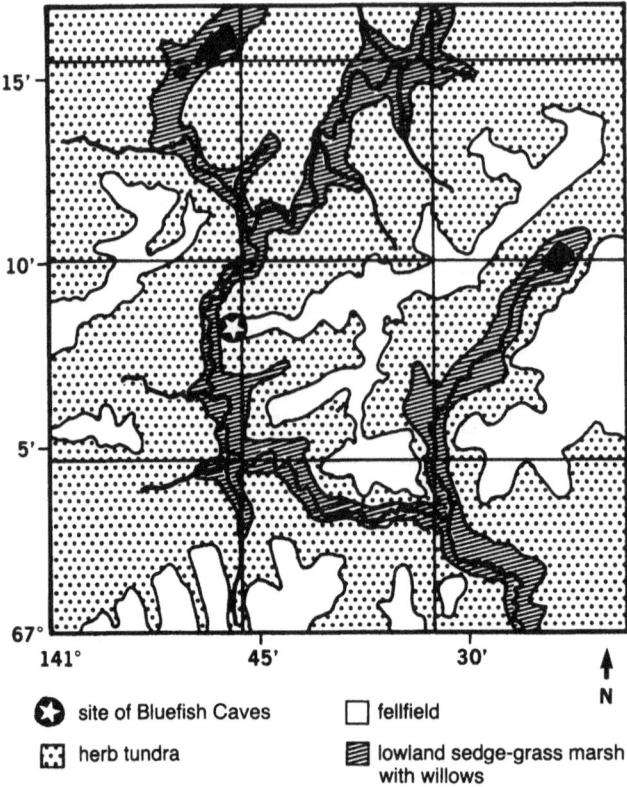

Figure 38 The late-glacial vegetation of the Bluefish River area, North Yukon, based on the pollen spectra in 15,500-yr-old Bluefish Cave sediments. Bluefish River flows from the southern edge of the map to the north; the lakes are Pear Lake in the northwest corner and Useful Lake. Scale 1:250,000; each square is 30 by 30 km

3. Individual mammal consumption rates for each community type are based on estimates of the proportions used by each species; the species totals are estimates of forage harvested annually. For example, mammoth is estimated to consume 29% of the willow forage, 21% of the sedge-grass marsh, and 9% of the sedge tundra for a total of 600 t (metric tonnes) yr^{-1} per 1,000 km^2.

4. It is then assumed that monogastric species consume 4% of their live body weight per day (2.2% for ruminants); therefore on an annual basis mammoth biomass, for example, is estimated at $600 / 1,460\% = 41$ t yr^{-1} per 1,000 km^2. The estimated body weight values for each species can then be used to derive estimates of the number of animals per unit area.

Now it is important to note that these rough calculations do no more than estimate the potential herbivore population, or carrying capacity, of a particular

Bluefish River

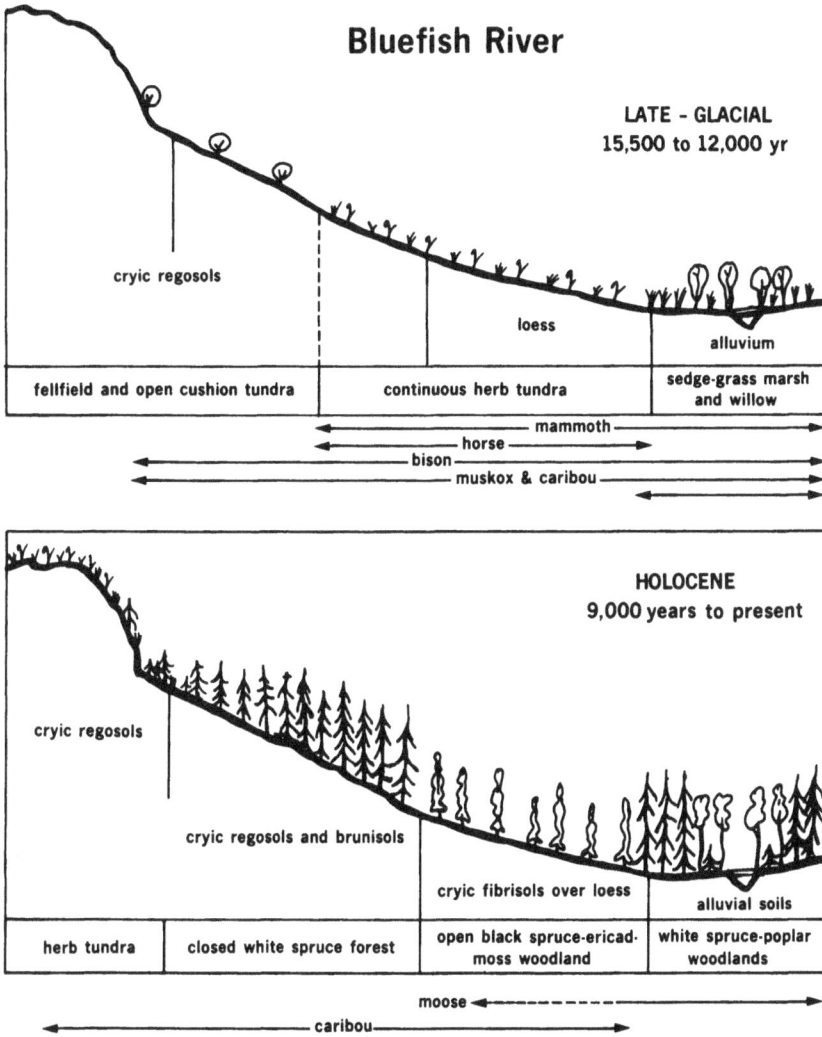

LATE - GLACIAL
15,500 to 12,000 yr

cryic regosols

loess

alluvium

| fellfield and open cushion tundra | continuous herb tundra | sedge-grass marsh and willow |

mammoth
horse
bison
muskox & caribou

HOLOCENE
9,000 years to present

cryic regosols

cryic regosols and brunisols

cryic fibrisols over loess

alluvial soils

| herb tundra | closed white spruce forest | open black spruce-ericad-moss woodland | white spruce-poplar woodlands |

moose
caribou

Figure 39 A schematic reconstruction of the late-glacial and Holocene landscapes of the Bluefish River watershed, showing the habitat utilization of the main large mammals

landscape complex, and are not a device to predict the actual numbers of animals that might have occupied the area at any moment in time past. No account is taken of such population regulatory mechanisms as predation, competition for habitat due to breeding behavioural traits, unfavourable climatic conditions, and other factors.

Table 17 An estimate of the herbivore carrying capacity of 1,000 km² surrounding the Bluefish Cave site, assuming a vegetation cover reconstructed from the late-glacial (12,500–15,500 yr BP) pollen data, as mapped in Figure 38. The assumptions and calculations are explained in the text.

		Lowland willow shrub	Lowland sedge-grass marsh	Upland herb tundra	Upland fell and cushion tundra
1	Area occupied (km²)	45	95	650	210
	Net above-ground primary production (t yr⁻¹ km⁻²)	200	50	10	5
	Net available production (t yr⁻¹ per 1,000 km²)	9,000	4,750	6,500	1,050
2	Net % harvested by herbivores	15	15	10	5
	Actual forage consumed (t yr⁻¹ per 1,000 km²)	1,350	713	650	52
3	Consumption by mammoth	(29%) 391	(21%) 149	(9%) 60	–
	horse	–	(10%) 71	(35%) 228	–
	bison	–	(16%) 114	(22%) 143	(50%) 26
	muskox	(14%) 189	(16%) 114	(17%) 110	(25%) 13
	caribou	(7%) 94	(21%) 150	(17%) 110	(25%) 13
	wapiti	(50%) 675	(16%) 114	–	–

		Mammoth	Horse	Bison	Muskox	Caribou	Wapiti
4	Forage harvested by species (t yr⁻¹ per 1,000 km²)	600	299	283	426	367	789
	Mean forage consumption as % of live weight	4	4	2.2	2.2	2.2	2.2
	Mean body weight (kg)	2,230	150	450	180	100	250
	Herbivore biomass (t per 1,000 km²)*	41	20	35	53	46	98
	Animal density (km⁻²)	0.018	0.13	0.08	0.29	0.46	0.39

*Based on annual food consumption of harvested forage per year.

However, this hypothetical reconstruction indicates that the topographically varied landscapes of the Bluefish River watershed and adjacent areas, with several rivers and streams, extensive valley lowlands with high water table, considerable relief (300–1,000 m over relatively short distances), slopes of varied steepness (15–35°) and aspect, and bedrock escarpments and ridges, could have supported a herbivore biomass of between 200 and 300 kg km⁻², which is slightly less than estimates for modern arctic landscapes (between 300 and 400 according to Guthrie 1968 and Hubert 1977), but about one-third of the value reached by Bliss and Richards (1982): 800–1,450 kg km⁻². I suggest that their estimates are too high, probably because of their assumption that 'the climate between 25,000 and 11,000 was warmer and drier in summer' than at present, and that the soil active layer was deeper than at present. Presumably these assumptions, which appear to be erroneous (CLIMAP 1976), led them to select the high values for net primary production.

The conclusion that unglaciated areas of North Yukon with adequate topographic diversity could have supported a fauna of six large herbivores remains simply a testable hypothesis. The botanical evidence supports the idea but requires amplification, particularly by more detailed pollen and macrofossil analyses of new sites. The reconstruction attempted above for the *late-glacial* in the Bluefish area of North Yukon suggests that small herds of low density of these species are not incompatible with the vegetation reconstruction developed from the pollen data. *Full-glacial* vegetation reconstructions based on pollen data (e.g. Cwynar 1982) would probably call for lower net primary production values than given in Table 17, and the proportion of the landscape occupied by continuous vegetation would have been smaller. But it is still feasible, as Cwynar concluded, 'that these herds may have been small with a low density of animals, as is the case today with muskox.'

The final question to be considered briefly in this section is: Does the palaeobotanical evidence contribute usefully to testing of the various hypotheses preferred to explain the extinction of several of the large mammal species?

The two classes of hypothesis are environmental change and cultural impact. The first was enunciated for northwest North America by Guthrie (1968) when he wrote (p 356): 'The change in Interior Alaska from grassland to coniferous forest and / or shrub tundra appears to have reduced the grazing habitat and increased interspecific competition until some species became extinct.' The other, often referred to as the 'prehistoric overkill model,' was developed by Martin (1967) and was summarized by him recently as follows (Martin 1982):

The global pattern of extinctions of all large mammals of the continents and islands appears to follow Paleolithic man's footsteps. Africa and southern parts of Asia are affected first with losses among the suids, primates, equids and giraffids over 40,000 years ago; Australia follows with extinction between 30,000 and 20,000 years ago of twelve genera of giant marsupials and size reduction of the surviving groups of large marsupials; between 20,000 and 10,000 years ago Europe and northern Asia are affected, to a relatively slight degree; North and South America are stripped of large herbivores between 12,000 and 10,000 years ago with only extinct species of *bison* persisting to the Holocene.

A variation or corollary of the cultural cause hypothesis is referred to by Guthrie (1968). It is that it was the full- and late-glacial predominance of a high-latitude herbivorous fauna that maintained the regional vegetation in a grassland rather than woodland or forest phase, much as domesticated grazing animals can repress forest encroachment of pastureland. The advent of prehistoric overkill removed this biotic factor and the vegetation succession proceeded to forest.

The botanical evidence that we have summarized schematically for the Bluefish River watershed (Figure 39, Table 18) confirms Guthrie's (1968) hypothesis that there was a sharp diminution in grazer-browser habitat at the beginning of the

Table 18 Estimated areas occupied by main vegetation types in a 1,000 km² block surrounding the Bluefish Caves site, for the late-glacial (12,500–15,500) and the Holocene (9,000 to present), showing also the rough proportions of the main topographic positions occupied by each type. The data refer to the area mapped in Figure 38.

	ALLUVIAL		LOWER SLOPE	UPPER SLOPE	RIDGE
Late Glacial					
Area occupied in km²	45	95	650		210
Vegetation type	sedge-grass marsh	willow shrub	herb tundra		fellfield and cushion tundra
Topographic position	ALLUVIAL		LOWER SLOPE	UPPER SLOPE	RIDGE
Holocene					
Vegetation type	white spruce-poplar–willow woodlands		black spruce-ericad-moss open woodlands	white spruce closed forests	herb tundra
Area occupied in km²	70		280	500	150

Holocene. In the mountainous and hilly terrain between roughly latitudes 66°N and 68°N in the North Yukon the proportion of the landscape covered by herbaceous, utilizable vegetation declined from about 75% to 15%, being replaced by cover dominated by plants that are either of low utilization value or unpalatable to all mammals: coniferous trees, ericoid shrubs, and *Sphagnum* moss. In other words suitable habitat for such grazers and grazer / browsers as mammoth, muskox, bison, and horse diminished by a factor of 5. In addition, the transition from a cold, continental late-glacial climate with annual snowfall <60 cm to Holocene conditions with snowfall values between 1.5 and 2.0 m probably contributed to making much of the North Yukon, at least south of the British Mountains, unsuitable for these animals. And no doubt that complex of poorly understood factors concerned with terrain surface conditions that limit the movement and mobility of herds changed at the beginning of the Holocene.

However, none of these changes is adequate as a cause of extinction since apparently suitable habitat for these arctic herbivores must have existed to the north, as it does today, on the arctic islands and in the northern mountains of the mainland.

The extinction hypotheses, if they are in fact testable, will yield primarily to secure archaeological and palaeontological data, possibly supplemented by vegetational reconstructions. A final point is perhaps worth making, however, that the major change in vegetation and therefore in potential habitat that we have documented for the transition from the latest glacial cycle to the current interglacial probably occurred many times during the Pleistocene at sites near the limits of forest and tundra, if we are justified in accepting the global extrapolations of the long, continuous marine and European terrestrial records (Kukla 1981). If the frequency of large-vertebrate extinction can be shown to be inordinately high in the late Pleistocene, then hypotheses of non-environmental causes are supported. At present the chronology of Beringian extinction is too imprecise to permit such a conclusion. This is true also for the rest of North America: Meltzer and Mead (1983) conclude from an analysis of 150 fossil sites in North America, excluding all Beringian sites, that 'while nearly 70% of all radiocarbon dates fall between 11,000 and 14,000 yr BP, this is largely due to sampling of the archaeological and geological records.'

2. Topographic Diversity and Palaeoecology

While many observers have recorded the great variation in plant cover at high latitudes in relation to differences of slope, aspect, exposure, and other elements of topography, few detailed investigations of the environment and vegetation of such features have been completed. Jacobs and Leung (1980) have provided some interesting data for a coastal area of Baffin Island, and they point out the importance for palaeoclimatic reconstructions of information on topoclimatic variation in relation to vegetation.

The most obvious example of these phenomena is the contrast in plant cover on slopes of different aspect near treeline. For example, in the area of our interest 20–30° slopes in the South Richardson Mountains have quite different plant communities than other sites have, apparently related primarily to aspect. As it happens, this area is one of the few to have escaped natural fires for at least a century, so that common source of vegetation variability can probably be discounted. The maximum solar angle at this latitude is 49°. North-facing slopes of long ridges usually support a herb-tundra sward and south-facing surfaces an open white spruce woodland, as in the example illustrated in Plate 24. Elsewhere in the same region (Plate 26) a third community type dominates relatively rare, steep slopes with northeast aspect, found in narrow, short valleys closed at their upper end by a high ridge. These slopes are dominated by a carpet of fruticose lichens (mainly *Cladonia* and *Cetraria*), with only scattered vascular plants, and the active layer in late summer is very shallow (20 cm). Apparently, and of course this is simply a hypothesis, the effect of a northeast aspect combined with the shading influence of the high ridge restricts the amount of solar radiation received to minimal values for the region. However, the relationship is probably not simple. In fact the maximum differences between solar radiation values on horizontal, north-facing 30°, and south-facing 30° surfaces at these latitudes do not occur during the growing season (mid-June to early August) but in April and May, as illustrated by values for Norman Wells, which lies just beyond the southeast corner of our area, and for Inuvik (Table 19, data made available by the Atmospheric Environment Service of Environment Canada). It is probable that the vegetation differences are due to several interrelated factors, including radiation differences, but also involving such winter conditions as exposure to wind and protection by snow cover.

In any event, it is likely that similar variation in vegetation cover over very short distances would have occurred in full- and late-glacial times under different macroclimatic regimes. Unfortunately our ability to develop palaeoecological reconstructions that take adequate account of microtopographic variation is hampered by the lack of modern ecological investigations.

It should also be pointed out here that the past vegetation of landscapes of considerable topographic and therefore ecological diversity is difficult to decipher using pollen and macrofossil analysis, both because average pollen dispersal distances greatly exceed the sizes of the different topographic and vegetation units and because upslope transport of pollen often blurs the registration of distinct communities (e.g. Maher 1972; Spear 1980). As a result we should be cautious in the conclusions we draw about past plant communities.

Investigations of the ecological factors controlling these major slope and aspect differences have not begun in northwest Canada, and have yet to produce conclusive results in adjacent Alaska (Slaughter and Long 1975) so we can engage only in hypothetical considerations at present.

Table 19 Mean daily global solar radiation (MJ m^{-2}) for horizontal, 30° north-facing, and 30° south-facing surfaces, at Norman Wells (65°N, 126°48′W) and Inuvik. Figures in parentheses are standard deviations. Data made available by the Atmospheric Environment Service, Environment Canada, Toronto

Surface station	April	May	June	July	August
0°					
NW	15.1 (4.1)	19.3 (5.7)	22.7 (6.4)	19.9 (5.3)	13.7 (5.3)
IN	15.6 (3.7)	20.1 (5.7)	22.6 (6.6)	19.7 (6.2)	12.2 (5.1)
30° S-facing)					
NW	20.3 (6.3)	21.5 (7.3)	23.9 (7.7)	21.7 (7.4)	16.5 (7.3)
IN	21.4 (6.1)	22.5 (7.7)	23.8 (7.7)	21.3 (7.5)	14.6 (7.1)
30° N-facing					
NW	7.2 (2.8)	13.8 (3.5)	18.3 (4.4)	15.5 (4.2)	8.5 (2.9)
IN	7.2 (3.2)	15.0 (3.4)	19.0 (4.7)	15.5 (4.4)	8.0 (2.7)

NW = Norman Wells, IN = Inuvik

It is likely that the following are, and have been in the past, important factors controlling the ecology of plant communities in regions of topographic diversity at high latitudes (Billings 1974; Dingman et al. 1980; among others):

1. Radiation – the amount of radiant energy at the ground surface, being the sum of incoming net radiation, net heat exchange with the atmosphere (sensible heat flux), net latent heat exchange with the atmosphere, and the net exchange of heat with soil and snow cover.

2. Water balance – the resultant of precipitation, net evapotranspiration, run-off, net infiltration, and changes in surface retention of water.

3. Snow cover – which varies according to amount and distribution of precipitation, redistribution by wind including accumulation and removal, and changes in density, crust characteristics, and resulting thermal properties.

4. Exposure to wind, ice, and sand-blasting – a function of elevation, aspect, and microtopography.

5. Soil instability due to frost thrusting and heaving, and other effects of the freeze-thaw cycle of soil moisture.

Now clearly these factors interact variably – with each other, and in relation to slope, aspect, surface roughness, geological materials, soil nutrients, and other factors – to determine the local habitat.

One of the most intriguing aspects of the effect of microclimatic factors on plant communities, but also one about which very little is known, is the effect of seasonality. Its potential importance becomes even greater when we recall that the astronomical theory of climatic change (Berger 1978, 1981) shows that seasonal differences in radiation input have varied from maximal values at ~10,000 yr BP to the present minimal values. In particular we should expect that, at the latitude of our particular area of study, the greatest differences between summer and winter radiation occurred in the early Holocene. We have adduced above certain evidence to support the hypothesis of maximum Holocene warming at 10,000–6,000 yr BP, but so far the detection of variation in pollen data attributable to seasonal change has not been possible. These tantalizing uncertainties lead us into a brief review of promising avenues for future investigation.

3. For the Future

Before pointing to possible lines for future investigation, let me emphasize that any palaeoecological conclusions I have drawn pertain only to northwest Canada and not to Beringia as a whole. As it is today, regional variation is likely to have been of great significance during the past. One example will suffice to illustrate. The vast lowlands of interior Alaska centred about the valleys of the Yukon and Tanana rivers immediately to the north of the Alaska Range have two important distinguishing features in their physical environment that undoubtedly were important also in the past: a characteristic climatic regime influenced by the proximity of the Pacific Ocean and the particular juxtaposition of mountain ranges; and a widespread mantle of thick loess that has been a factor of primary importance in determining soil properties. It is likely that these sharp contrasts with the area of our immediate interest had their full-glacial analogues, so it would be unwise to expect similar conditions to have prevailed in both regions.

The following are some of the more interesting lines of investigation that should attract our attention in the immediate future (I concede that my own scholarly predilections have undoubtedly influenced my selection):

1. Further pollen stratigraphic work should be restricted to areas where the record is sparse or lacking. In the northwest two such tracts of *terra incognita* exist. The first is the Hyndman and Anderson River uplands along with the intervening Inuvik-Anderson plains (Figure 7) that lie beyond the Tutsieta moraine (mapped in Figure 6). These old land surfaces have probably been free of continental glaciers since the maximum expansion of the Laurentide ice. Interstadial deposits from that time should exist in these areas, either in lake sediments or in exposures along stream channels. So far no such material has been found, but we plan to continue our efforts, so far unsuccessful, to core the several deep lakes in former glacial spillways, for example in the vicinity of Hyndman Lake (68°15′N, 131°10′W) and Peter Lake (68°45′N, 134°05′W). Recovery of primary limnic sediment from some of these sites might both resolve outstanding questions about the persistence of some plants in the area during the latest interstadial and help to clarify the chronology of the events of Pleistocene geology in the Lower Mackenzie region.

The second large gap is the North Yukon coastal lowlands. We have made preliminary soundings of several lakes that occupy glacial spillways on the mainland between Herschel Island and the Babbage River (Figure 12), and a few hold promise of yielding long records of sediment. That surface should be of the same age as the Hyndman-Anderson uplands, being within the extent of maximum Laurentide ice (Figure 5) but well beyond the limits of the final extension (Figure 6).

2. The most stimulating challenge for plant ecologists who use pollen analysis as their chief tool is to search for those sites and to employ those new methods that offer the possibility of bridging the wide gap between traditional vegetation history studies and plant population ecology. Closing the gap requires finer resolution in both time interval and geographical scale than we have traditionally used. The former can be achieved by the analysis of annually laminated lake sediments (see Saarnisto 1979 for a review of methods). They can provide short-interval (5–10 years) pollen influx data with accurate chronological control, amenable to powerful statistical analyses to detect possible cyclic processes in past plant communities, to test null hypotheses of random occurrence of sequence patterns, to apply phase and coherency analyses, and to sense trends in species interactions. Some of the directions of this exciting prospect have been set out in contributions by Watts (1973), Green (1981), Birks and Birks (1980), and Walker (1982a, b).

Improving the resolution of pollen analysis with respect to area representation can be accomplished by comparing pollen sequences from traditional repositories (e.g. lakes of area ~10 ha) with those from very small basins (<0.1 ha). The latter, particularly in wooded landscapes, give a more sensitive and detailed record of changes in the local vegetation surrounding the basin, while the proportion of regional, background pollen rain is greatly reduced. A particularly effective example of this approach is the investigation by Bradshaw (1981).

Both lakes with annually laminated sediments and small basins can be found in the northwest, though neither will turn out to be of common occurrence. They hold the key to our future attempts to elucidate in detail the rapid changes in plant community dynamics that occurred in the short interval between the latest glacial period and the present interglacial, when, as our existing data show, drastic alterations in the composition and structure of plant communities occurred.

3. Plant palaeoecologists will undoubtedly continue their efforts to work out in greater detail the floristic composition and therefore palaeoenvironmental indications of the full-glacial assemblages registered as both pollen and macrofossils in primary lake sediment. The alternative hypotheses of full-glacial landscapes require extensive detailed testing against such information.

4. I have already, at the end of Chapter 6, referred to an active and fruitful avenue of palaeoenvironmental research that will become important in the immediate future: the application of calibration functions derived from large sets of modern pollen and climate data to sub-fossil pollen assemblages (Webb and Clark 1977; Arigo et al. 1982). The resulting palaeoclimatic estimates can be used to test conclusions drawn from the plant indicator method, and can themselves be tested with independent data sets such as ostracod analyses. Recently Howe and

Webb (1983) have made notable advances in the methodology of calibration functions, answering several of the points of reservation and criticism raised by Birks (1981). Interested readers can find a stimulating review of the current status of this and other aspects of Quaternary palaeoecology in an essay by Prentice (1983).

5. The information at hand provides hints that during the early Holocene, particularly in landscapes with extensive lowlands, major changes occurred in the permafrost characteristics, hydrology, and resulting water chemistry of watersheds in response to paludification. It is likely that there were attendant changes in the thickness of the active layer; in the development of palsas, hummocks, and other microtopographic features; in the plant cover, such as replacement of white spruce–willow–herb communities by black spruce–ericad–*Sphagnum* woodlands; in the development of ice-wedge polygons in the northern region; and in changes in the chemical regime of lakes and ponds as registered in the sediments. The broader implication of these changes bears on questions of available habitat for major elements of the fauna. Detailed studies of suitable sites in both the northern, continuous permafrost zone (already in hand in the Old Crow Flats by Ovenden 1981 and current work) and in the extensive treed mires of the Mackenzie Valley–Peel Plain Lowlands region would clarify and elaborate these early intimations.

6. Part of the reason for the development of the new trends in plant palaeoecology, and particularly in pollen analysis, is that the enormous volume of statistics produced by several thoroughly analysed sections of late Quaternary sediment can be stored and manipulated easily by modern computer methods. In a similar way efficient data-processing has expanded the scope of simulation methods in plant ecology. At least two types offer promise to the plant palaeoecologist.

The first, so far as I know not yet attempted, is the plant productivity simulation procedures (so-called models) developed for agricultural crops, particularly to predict crop yields in response to varied weather conditions. In a recent assessment of the great torrent of such models, Monteith (1981) has warned that many have become too complex, including large numbers of variables, often lacking reliable values because of the absence of experimental data. The late G. Szeicz and I made a start recently with the application of Monteith's (1966) photosynthesis model originally developed for crops, and Penman's (1948) potential evaporation equations, to estimate the potential growth (production) of perennial plants under subarctic and arctic climatic regimes. The basic data entered for each site are: latitude; leaf area index; several optional values for photosynthetic efficiency; mean, maximum, and minimum daily temperature; relative humidity; precipitation; wind speed; and mean daily global solar radiation calculated for horizontal ground and 30° north-facing and 30° south-facing slopes. The ratio of precipitation to potential evaporation is calculated by Penman's (1948) equations. Threshold minimal temperatures and water balance values can be chosen below which growth ceases, based on data from the literature on arctic ecology (e.g. review by Billings 1974). Now, as Monteith (1981) concedes, 'there are many objections to this procedure': it overlooks the correlation and interaction of factors, the adaptation of plants to weather stress, restrictions due to soil nutrients, and many biotic factors. However, as he concludes: 'Despite all these limitations I believe it is worth making the effort to distinguish the main climate-dependent mechanisms which determine growth and yield from those which play a secondary role.' Our preliminary results have produced simulated growth (annual net production) values for Inuvik, Tuktoyaktuk, and Sachs Harbour that are very similar to actual values for comparable stations in the north. However, we have not yet adapted the calculations to take adequate account of the effect of permafrost melt in spring, which releases available moisture in the soil independently of atmospheric moisture. The effort continues. The ultimate objective, if sensible simulations can be derived for modern situations, is to estimate full- and late-glacial climatic conditions from the fossil pollen record, as we have done in Chapter 6, and then derive estimates of potential production for different habitats in a topographically diverse landscape. These results should contribute towards a solution of the problems of reconciling a diverse full- and late-glacial fauna with the reconstructed vegetation.

The other type of simulation study uses basic information on the autecology of the chief species involved in an area to generate hypothetical plant communities, expressed as estimates of species abundance. So far a model using the chief tree species of the forests of the eastern United States has been used by Solomon et al. (1980) to simulate 16,000 years of forest history in Tennessee, as a basis for comparison with the pollen record. The simulation produces change in forest composition only as a function of the biological characteristics of the species – such variables as seed production, rate of growth longevity, shade and moisture tolerance, and average maximum height. The value of this approach is that hypotheses relating climate control to biotic factors such as competition can be tested. Special problems will present themselves when these methods are attempted with high-latitude communities, but they are not beyond resolution.

A common *sine qua non* of many of the above lines of future research is a secure foundation of information on the autecology of the more important plants. We need many more precise, experimental studies of the factors that control the reproduction and growth of the taxa that recur in our sub-fossil assemblages. Otherwise our new analytical tools will not help us much.

And of course the continued progress of plant palaeoecology, as for all historical sciences, will always lean heavily on the products of serendipity.

APPENDIXES

APPENDIX 1

A listing in taxonomic order of the species of vascular plants that have been recorded for the study area

OPHIOGLOSSACEAE
Botrychium boreale
  ssp. *obtusilobum*
Botrychium lunaria

POLYPODIACEAE
Cryptogramma stelleri
Cystopteris fragilis
Cystopteris montana
Dryopteris fragans
Woodsia alpina
Woodsia glabella
Woodsia ilvensis

EQUISETACEAE
Equisetum arvense
Equisetum fluviatile
Equisetum palustre
Equisetum pratense
Equisetum scirpoides
Equisetum sylvaticum
  var. *pauciramosum*
Equisetum variegatum

LYCOPODIACEAE
Lycopodium alpinum
Lycopodium annotinum
Lycopodium
  complanatum
Lycopodium selago

SELAGINELLACEAE
Selaginella sibirica

PINACEAE
Juniperus communis

Juniperus horizontalis
Larix laricina
Picea glauca
Picea mariana

SPARGANIACEAE
Sparganium
  angustifolium
Sparganium
  hyperboreum
Sparganium minimum
Sparganium
  multipedunculatum

POTAMOGETONACEAE
Potamogeton alpinus
  ssp. *tenuifolius*
Potamogeton filiformis
Potamogeton friesii
Potamogeton
  gramineus
Potamogeton
  pectinatus
Potamogeton
  porsildiorum
Potamogeton
  praelongus
Potamogeton
  richardsonii
Potamogeton
  strictifolius var.
  rutiloides
Potamogeton vaginatus
Potamogeton
  zosteriformis

SCHEUCHZERIACEAE
Triglochin maritimum
Triglochin palustre

GRAMINEAE
Agropyron sericeum
Agropyron
  trachycaulum
Agropyron violaceum
  ssp. *violaceum*
Agrostis borealis
Agrostis scabra
Alopecurus alpinus
Arctagrostis
  arundinacea var.
  arundinacea
Arctagrostis latifolia
  ssp. *latifolia*
Arctophila fulva
Beckmannia
  syzigachne
Bromus pumpellianus
  var. *pumpellianus*
Bromus pumpellianus
  var. *arcticus*
Calamagrostis
  canadensis var.
  langsdorffii
Calamagrostis
  chordorrhiza
Calamagrostis
  deschampsioides
Calamagrostis
  inexpansa

*Calamagrostis
    lapponica* var.
    nearctica
*Calamagrostis
    purpurascens*
Deschampsia brevifolia
*Deschampsia
    caespitosa*
Dupontia fisheri ssp.
    fisheri
Dupontia fisheri ssp.
    psilosantha
Elymus arenarius ssp.
    mollis
Elymus innovatus
Festuca altaica
Festuca baffinensis
Festuca brachyphylla
Festuca rubra
Hierochloë alpina
Hierochloë odorata
Hierochloë pauciflora
Hordeum jubatum
Koeleria asiatica
Phippsia algida
Poa alpigena var.
    alpigena
Poa alpina
Poa ammophila
Poa arctica ssp. *arctica*
Poa arctica ssp.
    caespitans
Poa glauca
Poa lanata
Poa paucispicula
Puccinellia agrostidea
Puccinellia andersonii
Pucinellia arctica
Puccinellia borealis
Puccinellia contracta
*Puccinellia
    phryganodes*
Puccinellia vaginata
Trisetum sibiricum
Trisetum spicatum s.
    lat.

CYPERACEAE
Carex albo-nigra
Carex amblyorhyncha

Carex aquatilis var.
    aquatilis
Carex aquatilis var.
    stans
Carex atrofusca
Carex aurea
Carex bicolor
Carex bonanzensis
Carex canescens
Carex capillaris ssp.
    capillaris
Carex capillaris ssp.
    chlorostachys
Carex capillaris ssp.
    robustior
Carex capitata
Carex chordorrhiza
Carex concinna
Carex consimilis
Carex diandra
Carex disperma
Carex eburnea
Carex eleusinoides
Carex garberi
Carex glacialis
Carex glareosa var.
    amphigena
Carex gynocrates
Carex holostoma
Carex lachenalii
Carex lapponica
Carex laxa
Carex limosa
Carex livida var.
    grayana
Carex lugens
Carex mackenziei
Carex macloviana
Carex maritima
Carex media
Carex membranacea
Carex microchaeta
Carex microglochin
Carex misandra
Carex morrisseyi
Carex nardina
Carex obtusata
Carex paupercula
Carex petricosa
Carex physocarpa
Carex podocarpa

Carex ramenskii
Carex rariflora var.
    rariflora
Carex rostrata
Carex rotundata
Carex rupestris
Carex scirpoidea
Carex subspathacea
Carex supina ssp.
    spaniocarpa
Carex tenuiflora
Carex ursina
Carex vaginata
Carex williamsii
Eleocharis acicularis
Eleocharis palustris
*Eriophorum
    angustifolium*
*Eriophorum
    brachyantherum*
Eriophorum callitrix
Eriophorum russeolum
    var. *albidum*
*Eriophorum
    scheuchzeri*
Eriophorum triste
Eriophorum vaginatum
    ssp. *vaginatum*
Kobresia hyperborea
Kobresia myosuroides
*Kobresia
    simpliciuscula*
Scirpus caespitosus
    ssp. *austriacus*

ARACEAE
Calla palustris

LEMNACEAE
Lemna trisulca

JUNCACEAE
Juncus albescens
Juncus alpinus ssp.
    nodulosus
Juncus balticus var.
    alaskanus
Juncus biglumis
Juncus bufonius
Juncus castaneus
Luzula arcuata

Luzula confusa
Luzula groenlandica
Luzula multiflora ssp.
  frigida var.
  contracta
Luzula nivalis var.
  nivalis
Luzula nivalis var.
  latifolia
Luzula parviflora
Luzula rufescens
Luzula wahlenbergii

LILIACEAE
Allium schoenoprasum
  var. *sibiricum*
Lloydia serotina
Tofieldia coccinea
Tofieldia glutinosa
Tofieldia pusilla
Zygadenus elegans

ORCHIDACEAE
Corallorhiza trifida
Cypripedium
  guttatum
Cypripedium
  passerinum
Habenaria obtusata
Listera borealis
Orchis rotundifolia

SALICACEAE
Populus balsamifera
Populus tremuloides
Salix alaxensis
Salix arbusculoides
Salix arctica
Salix arctophila
Salix barclayi
Salix barrattiana
Salix bebbiana
Salix brachycarpa
Salix chamissonis
Salix dodgeana
Salix farriae
Salix fuscescens
Salix glauca var.
  glauca
Salix glauca var.
  acutifolia

Salix glauca var.
  aliceae
Salix glauca var.
  stenolepis
Salix interior var.
  pedicellata
Salix lanata ssp.
  Richardsonii
Salix longistylis
Salix myrtillifolia
Salix niphoclada
Salix ovalifolia var.
  arctolitoralis
Salix phlebophylla
Salix planifolia
Salix polaris ssp.
  pseudopolaris
Salix pulchra
Salix reticulata

MYRICACEAE
Myrica gale

BETULACEAE
Alnus crispa
Alnus incana ssp.
  tenuifolia
Betula glandulosa
Betula occidentalis
Betula papyrifera var.
  neoalaskana

SANTALACEAE
Geocaulon lividum

POLYGONACEAE
Koenigia islandica
Oxyria digyna
Polygonum alaskanum
Polygonum amphibium
  var. *stipulaceum*
Polygonum bistorta
  ssp. *plumosum*
Polygonum caurianum
Polygonum viviparum
Rumex acetosa ssp.
  alpestris
Rumex arcticus
Rumex triangulivalvis

CHENOPODIACEAE
Atriplex gmelinii
Chenopodium
  berlandieri var.
  zschackei
Chenopodium
  capitatum
Chenopodium glaucum
  ssp. *salinum*
Monolepis nuttalliana

PORTULACACEAE
Claytonia sarmentosa
Claytonia tuberosa
Montia lamprosperma

CARYOPHYLLACEAE
Arenaria capillaris var.
  nardifolia
Arenaria humifusa
Cerastium arvense
Cerastium
  beeringianum
Cerastium maximum
Cerastium regelii
Dianthus repens
Honckenya peploides
  var. *diffusa*
Melandrium affine
Melandrium apetalum
  ssp. *arcticum*
Melandrium ostenfeldii
Melandrium
  taimyrense
Melandrium taylorae
Minuartia arctica
Minuartia biflora
Minuartia dawsonensis
Minuartia macrocarpa
Minuartia rossii
Minuartia rubella
Minuartia yukonensis
Moehringia lateriflora
Sagina intermedia
Silene acaulis ssp.
  acaulis
Silene acaulis ssp.
  subacaulescens
Silene repens ssp.
  purpurata
Stellaria calycantha

Stellaria crassifolia
Stellaria edwardsii
Stellaria humifusa
Stellaria laeta
Stellaria longipes
Stellaria monantha
Stellaria stricta
Stellaria subvestita
Wilhelmsia physodes

RANUNCULACEAE
Aconitum
  delphinifolium
Anemone drummondii
Anemone multiceps
Anemone multifida
Anemone narcissiflora
Anemone parviflora
Anemone richardsonii
Aquilegia brevistyla
Caltha palustris var.
  palustris
Caltha palustris var.
  arctica
Delphinium
  brachycentrum
Delphinium glaucum
Pulsatilla ludoviciana
Ranunculus aquatilis
  var. *eradicatus*
Ranunculus aquatilis
  var. *subrigidus*
Ranunculus cymbalaria
Ranunculus
  eschscholtzii
Ranunculus flammula
Ranunculus gelidus
Ranunculus gmelinii
Ranunculus
  hyperboreus
Ranunculus lapponicus
Ranunculus macounii
Ranunculus nivalis
Ranunculus pallasii
Ranunculus pedatifidus
  var. *leiocarpus*
Ranunculus pygmaeus
Ranunculus sabinei
Ranunculus sceleratus
  ssp. *multifidus*
Ranunculus sulphureus

Ranunculus turneri
Thalictrum alpinum
Thalictrum venulosum

FUMARIACEAE
Corydalis pauciflora
Corydalis sempervirens

CRUCIFERAE
Alyssum americanum
Arabis divaricarpa
Arabis drummondii
Arabis hirsuta ssp.
  pynocarpa
Arabis lyrata var.
  kamchatica
Barbarea orthoceras
Braya glabella
Braya henryae
Braya humilis s. lat.
Braya purpurascens
Cardamine bellidifolia
Cardamine digitata
Cardamine
  microphylla
Cardamine pratensis
  var. *angustifolia*
Cochlearia officinalis
Descurainia sophioides
Draba alpina
Draba aurea
Draba cinerea
Draba corymbosa
Draba crassifolia
Draba fladnizensis
Draba glabella
Draba incerta
Draba lactea
Draba longipes
Draba nivalis
Draba oliogosperma
Draba palanderiana
Draba pilosa
Erysimum
  cheiranthoides
Erysimum
  inconspicuum
Erysimum pallasii
Eutrema edwardsii
Halimolobos mollis
Lesquerella arctica

Lesquerella calderi
Parrya arctica
Parrya nudicaulis
Rorippa barbareaefolia
Rorippa calycina
Rorippa islandica
Smelowskia calycina
  var. *media*
Thellungiella
  salsuginea
Thlaspi arcticum

DROSERACEAE
Drosera anglica
Drosera rotundifolia

CRASSULACEAE
Rhodiola integrifolia

SAXIFRAGACEAE
Boykinia richardsonii
Chrysosplenium
  tetrandrum
Chrysosplenium
  wrightii
Parnassia kotzebuei
Parnassia palustris var.
  neogaea
Ribes hudsonianum
Ribes triste
Saxifraga bronchialis
  ssp. *Funstonii*
Saxifraga caespitosa
  ssp. *caespitosa*
Saxifraga caespitosa
  ssp. *monticola*
Saxifraga caespitosa
  ssp. *uniflora*
Saxifraga cernua
Saxifraga davurica ssp.
  grandipetala
Saxifraga eschscholtzii
Saxifraga ferruginea
Saxifraga flagellaris
  ssp. *flagellaris*
Saxifraga foliolosa
Saxifraga hieracifolia
Saxifraga hirculus s.
  lat.
Saxifraga nivalis
Saxifraga oppositifolia

Saxifraga punctata
   ssp. *Nelsoniana*
Saxifraga radiata
Saxifraga reflexa
Saxifraga rivularis
Saxifraga serpyllifolia
Saxifraga tricuspidata

ROSACEAE
Dryas alaskensis
Dryas crenulata
Dryas drummondii
Dryas integrifolia
Dryas octopetala
Dryas punctata
Dryas sylvatica
Fragaria virginiana
   ssp. *glauca*
Geum glaciale
Geum rossii
Luetkea pectinata
Potentilla anserina
Potentilla biflora
Potentilla egedii
Potentilla elegans
Potentilla fruticosa
Potentilla hyparctica
   var. *elatior*
Potentilla multifida
Potentilla nivea ssp.
   nivea
Potentilla nivea ssp.
   Chamissonis
Potentilla nivea ssp.
   Hookeriana
Potentilla norvegica
Potentilla palustris
Potentilla pensylvanica
Potentilla pulchella s.
   lat.
Potentilla rubricaulis
Potentilla uniflora
Potentilla vahliana
Rosa acicularis
Rubus acaulis
Rubus arcticus
Rubus chamaemorus
Rubus pubescens
Rubus strigosus
Sanguisorba officinalis

Sibbaldia procumbens
Spiraea beauverdiana

LEGUMINOSAE
Astragalus alpinus
Astragalus bodinii
Astragalus eucosmus
Astragalus richardsonii
Astragalus tenellus
Astragalus umbellatus
Hedysarum alpinum
   var. *americanum*
Hedysarum mackenzii
Lathyrus japonicus var.
   aleuticus
Lupinus arcticus
Oxytropis arctica
Oxytropis deflexa var.
   foliolosa
Oxytropis glutinosa
Oxytropis hyperborea
Oxytropis jordalii
Oxytropis maydelliana
   s. lat.
Oxytropis nigrescens
   ssp. bryophylla
Oxytropis nigrescens
   ssp. pygmaea
Vicia americana

LINACEAE
Linum lewisii

CALLITRICHACEAE
Callitriche
   hermaphroditica
Callitriche verna

EMPETRACEAE
Empetrum nigrum ssp.
   hermaphroditum

VIOLACEAE
Viola epipsila ssp.
   repens

ELAEAGNACEAE
Elaeagnus commutata
Shepherdia canadensis

ONAGRACEAE
Epilobium
   angustifolium
Epilobium arcticum
Epilobium davuricum
Epilobium latifolium
Epilobium palustre

HALORAGACEAE
Hippuris tetraphylla
Hippuris vulgaris
Myriophyllum
   exalbescens
Myriophyllum
   verticillatum var.
   pectinatum

UMBELLIFERAE
Bupleurum
   americanum
Cicuta mackenzieana
Conioselinum
   cnidiifolium

CORNACEAE
Cornus stolonifera

PYROLACEAE
Moneses uniflora
Pyrola chlorantha
Pyrola grandiflora
Pyrola minor
Pyrola secunda s. lat.

ERICACEAE
Andromeda polifolia
Arctostaphylos alpina
Arctostaphylos rubra
Arctostaphylos uva-
   ursi
Cassiope tetragona
   ssp. *tetragona*
Chamaedaphne
   calyculata
Ledum decumbens
Ledum groenlandicum
Loiseleuria
   procumbens
Oxycoccus
   microcarpus

*Rhododendron
  lapponicum*
Vaccinium uliginosum
  s. lat.
Vaccinium vitis-idaea
  var. *minus*

DIAPENSIACEAE
Diapensia obovata

PRIMULACEAE
*Androsace
  chamaejasme* var.
  arctica
*Androsace
  septentrionalis*
Dodecatheon frigidum
Douglasia arctica
Douglasia ochotensis
Lysimachia thyrsiflora
Primula borealis
Primula egaliksensis
Primula stricta
*Primula
  tschuktschorum* ssp.
  Cairnesiana

PLUMBAGINACEAE
Armeria maritima ssp.
  arctica
Armeria maritima ssp.
  labradorica

GENTIANACEAE
Gentiana arctophila
Gentiana glauca
Gentiana propinqua
Gentiana prostrata
Gentiana raupii
Gentiana richardsonii
*Lomatogonium
  rotatum* ssp. *rotatum*

POLEMONIACEAE
Phlox alaskensis
Phlox richardsonii
*Polemonium
  acutiflorum*
Polemonium boreale
*Polemonium
  pulcherrimum*

BORAGINACEAE
Eritrichium aretioides
Eritrichium splendens
Lappula redowskii ssp.
  occidentalis
Mertensia maritima
Mertensia paniculata
  var. *paniculata*
Myosotis alpestris ssp.
  asiatica

LABIATAE
Stachys palustris

SCROPHULARIACEAE
Castilleja caudata
Castilleja elegans
Castilleja hyperborea
Castilleja raupii
Castilleja yukonis
Lagotis stelleri
Pedicularis arctica
Pedicularis capitata
Pedicularis labradorica
Pedicularis lanata
Pedicularis lapponica
Pedicularis sudetica
Pedicularis verticillata
Synthyris borealis
Veronica wormskjoldii
  ssp. *wormskjoldii*

OROBANCHACEAE
Boschniakia rossica

LENTIBULARIACEAE
Pinguicula villosa
Pinguicula vulgaris
Utricularia intermedia
Utricularia ochroleuca
Utricularia vulgaris

PLANTAGINACEAE
Plantago canescens
Plantago eriopoda
Plantago major

RUBIACEAE
Galium boreale
Galium trifidum

CAPRIFOLIACEAE
Linnaea borealis var.
  americana
Viburnum edule

VALERIANACEAE
Valeriana capitata

CAMPANULACEAE
Campanula aurita
Campanula lasiocarpa
Campanula uniflora

COMPOSITAE
Achillea nigrescens
Achillea sibirica
Antennaria angustata
Antennaria compacta
Antennaria crymophila
Antennaria ekmaniana
Antennaria isolepis
*Antennaria
  monocephala*
*Antennaria
  neoalaskana*
Antennaria oxyphylla
*Antennaria
  pedunculata*
Antennaria philonipha
Antennaria pulcherrina
Antennaria pygmaea
*Antennaria
  subcanescens*
Arnica alpina ssp.
  angustifolia
Arnica alpina ssp.
  attenuata
Arnica alpina ssp.
  tomentosa
Arnica lessingii
Arnica lonchophylla
Arnica louiseana ssp.
  frigida
Artemisia alaskana
Artemisia arctica ssp.
  arctica
Artemisia borealis
Artemisia canadensis
Artemisia frigida
Artemisia globularia
Artemisia glomerata

Artemisia hyperborea
Artemisia tilesii
Aster alpinus ssp.
  vierhapperi
Aster pygmaeus
Aster sibiricus
Chrysanthemum
  arcticum
Chrysanthemum
  integrifolium
Crepis elegans
Crepis nana
Erigeron acris var.
  asteroides
Erigeron compositus s.
  lat.
Erigeron elatus
Erigeron eriocephalus
Erigeron glabellus ssp.
  pubescens
Erigeron grandiflorus
  ssp. *arcticus*

Erigeron humilis
Erigeron hyssopifolius
Erigeron philadelphicus
Erigeron purpuratus
Erigeron yukonensis
Matricaria ambigua
Petasites arcticus
Petasites frigidus
Petasites hyperboreus
Petasites sagittatus
Saussurea angustifolia
  var. *angustifolia*
Saussurea angustifolia
  var. *yukonensis*
Senecio atropurpureus
Senecio congestus
Senecio cymbalaria
Senecio hyperborealis
Senecio kjellmanii
Senecio lindstroemii

Senecio lugens
Senecio yukonensis
Solidago decumbens
  var. *oreophila*
Solidago multiradiata
Tanacetum huronense
Taraxacum alaskanum
Taraxacum dumetorum
Taraxacum integratum
Taraxacum lacerum
Taxaracum
  mackenziense
Taraxacum
  maurolepium
Taraxacum pellianum
Taraxacum
  phymatocarpum
Taraxacum pumilum
Taraxacum sibiricum

APPENDIX 2

Tabular summaries of the vegetation composition of the main community types

(a) Main species and their Domin cover values in the ground vegetation of white spruce stands on recent alluvium; data from Gill (1973c) and my own records

	Stands									
	1	2	3	4	5	6	7	8	9	$\overline{X}$
SHRUB LAYER (0.5–3m)										
Alnus crispa	5	3	3	4	3	4	4	5	4	4
Salix alaxensis	1	2	1	–	1	1	2	1	1	1
S. arbusculoides	2	3	3	2	4	1	4	3	2	3
S. glauca	3	4	4	1	3	1	4	2	3	3
S. richardsonii	–	–	1	–	1	–	–	–	1	0.3
S. barclayi	–	–	–	1	–	–	1	–	1	0.3
Shepherdia canadensis	–	–	–	1	–	2	1	1	1	0.7
Cornus stolonifera	–	–	–	1	1	2	1	1	1	0.8
Populus balsamifera	1	–	1	–	–	1	1	–	–	0.3
HERB-MOSS LAYER										
Arctostaphylos rubra	4	5	3	4	4	5	4	5	5	4
Hedysarum alpinum	2	1	2	3	1	1	2	1	1	2
Rosa acicularis	1	2	2	2	1	2	2	1	1	2
Pyrola grandiflora	2	1	2	1	1	1	1	2	1	1
P. minor	–	–	–	1	–	1	1	2	1	0.7
Moneses uniflora	1	–	1	–	1	1	1	–	1	0.7
Listera borealis	–	–	1	–	1	–	–	–	–	0.2
Rubus arcticus	2	1	1	1	1	2	–	1	1	1
Boschniakia rossica	1	1	1	2	1	1	–	1	1	1
Equisetum arvense	2	2	1	2	2	1	1	1	1	1
Habenaria obtusata	1	–	1	1	–	–	–	1	1	0.5
Aster sibiricus	1	1	–	–	1	1	1	1	1	1
Anemone parviflora	–	–	–	–	1	1	–	1	–	0.3
Delphinium glaucum	–	–	–	–	–	1	1	–	–	0.2
Tomenthypnum nitens	4	3	4	3	3	3	4	4	4	4
Hylocomium alaskanum	2	3	3	3	2	2	2	3	2	3
Campylium stellatum	2	1	2	2	1	2	2	1	1	2
Distichium capillaceum	1	1	2	1	1	2	1	1	1	1
Drepanocladus uncinatus	2	3	2	2	1	2	2	2	1	2
Aulacomnium palustre	1	2	1	1	1	2	1	1	1	1
Eurhynchium pulchellum	1	1	1	1	2	1	1	1	1	1

(b) Main species and their Domin cover values in the ground
vegetation of white spruce stands on uplands with non-calcareous
parent materials; data from Inglis (1975) and my own records

	Stands							$\bar{X}$
Vaccinium vitis-idaea	5	5	6	6	3	5	5	5
Empetrum nigrum	4	3	5	2	4	3	3	3.4
Rosa acicularis	1	4	3	2	3	3	5	3
Ledum decumbens	6	3	6	7	–	3	3	4
Betula glandulosa	5	3	2	5	–	3	2	2.9
Salix glauca	2	4	4	4	3	–	3	2.9
Arctagrostis latifolia	2	–	3	1	1	2	4	1.9
Pyrola secunda	1	1	3	1	1	–	–	1
Alnus crispa	2	–	5	3	1	–	–	1.6
Linnaea borealis	–	4	–	–	5	–	2	1.6
Juniperus communis	–	4	–	–	4	–	4	1.7
Lupinus arcticus	3	2	–	–	–	–	4	1.3
Arctostaphylos rubra	–	1	3	–	4	–	–	1.1
Ribes triste	–	–	3	–	2	–	4	1.3
Petasites frigidus	3	–	–	2	–	3	–	1.1
Vaccinium uliginosum	3	–	3	–	–	2	–	1.1
Rubus chamaemorus	3	–	–	3	–	2	–	1.1
Carex lugens	1	–	–	–	2	–	3	0.9
Equisetum arvense	2	–	–	–	–	1	1	0.6
Arctostaphylos uva-ursi	3	5	–	–	–	–	–	1.1
Shepherdia canadensis	–	1	–	–	3	–	–	0.6
Epilobium angustifolium	–	3	–	–	–	–	2	0.7
Equisetum scirpoides	–	–	3	–	1	–	2	0.9
Hylocomium splendens	1	1	4	3	4	1	5	2.7
Polytrichum strictum	1	3	1	2	2	4	3	2.3
Ptilidium ciliare	1	–	5	4	3	4	3	2.9
Dicranum elongatum	3	2	3	5	–	1	3	2.4
Aulacomnium turgidum	–	3	3	4	2	–	2	2
Hypnum sp.	–	1	3	1	4	–	1	1.4
Drepanocladus uncinatus	2	1	2	–	2	–	3	1.4
Dicranum sp.	–	–	3	–	4	3	–	1.4
Mnium sp.	3	–	–	4	1	–	–	1.1
Aulacomnium palustre	4	–	–	1	2	–	–	1
Distichium capillaceum	–	–	2	–	2	–	3	1
Tomenthypnum nitens	1	–	2	–	2	–	–	0.7
Sphagnum girgensohnii	4	–	–	4	–	–	–	1.1
Sphenobolus minutus	3	–	–	2	–	–	–	0.7
Cladonia mitis	4	2	5	7	3	6	3	4.3
Peltigera aphthosa	5	1	4	4	5	2	5	3.7
Cetraria pinastri	1	1	1	1	1	1	1	1
Hypogymnia physodes	1	1	1	1	1	1	–	0.9
Cladonia rangiferina	4	–	4	5	1	5	1	2.8
Cetraria nivalis	1	2	–	2	1	5	5	2.3
Cladonia gracilis	4	1	3	–	3	2	2	2.1
Cetraria sepincola	1	1	1	–	1	–	1	0.7
Peltigera malacea	3	3	–	–	3	–	3	1.7

	Stands							$\overline{X}$
Peltigera canina	–	2	1	1	4	–	–	1.1
Cladonia chlorophaea	–	1	2	–	2	1	–	0.9
Cladonia uncialis	–	1	2	2	–	1	–	0.9
Parmeliopsis ambigua	1	1	1	–	1	–	–	0.6
Alectoria lanestris	–	–	1	–	1	1	1	0.6
Stereocaulon sp.	1	2	–	–	–	5	–	1.1
Cladonia cornuta	–	–	1	3	–	1	–	0.7
Cladonia fimbriata	2	1	–	–	–	–	1	0.6
Parmelia sulcata	–	–	1	–	2	–	1	0.6
Cladonia amaurocraea	1	–	–	–	1	1	–	0.4
Cladonia alpestris	–	–	–	3	–	3	–	0.9

(c) Main species and their Domin cover values in the ground vegetation of white spruce stands on uplands with calareous parent materials and northern aspect; data from Ritchie (1982)

	Stands										$\overline{X}$
Dryas octopetala	4	3	4	3	3	2	4	4	3	2	3.2
Arctostaphylos rubra	2	–	3	2	2	2	3	2	2	2	2.1
Cassiope tetragona	3	4	3	2	–	3	–	3	–	3	2.1
Astragalus umbellatus	2	1	2	2	2	2	2	2	2	2	1.9
Carex scirpoidea	2	2	1	–	2	2	–	2	2	2	1.5
Oxytropis maydelliana	1	–	1	–	2	–	–	–	–	–	0.4
Salix reticulata	2	2	1	2	2	2	–	2	2	2	1.7
S. glauca	–	–	–	–	–	–	3	–	3	–	0.6
Pedicularis capitata	2	–	–	2	1	–	–	2	2	–	0.9
P. labradorica	–	2	–	–	–	2	–	–	–	2	0.6
Tofieldia pusilla	–	1	–	2	–	–	2	2	–	2	0.9
Polygonum viviparum	–	–	1	–	2	–	–	–	–	–	0.3
Saxifraga oppositifolia	–	–	–	–	–	–	2	–	–	–	0.2
Carex capillaris	2	2	–	2	–	2	–	2	1	2	1.3
Hedysarum alpinum	–	–	3	–	3	–	–	–	3	–	0.9
Lupinus arcticus	–	–	2	3	–	–	3	–	–	2	1
Papaver macounii	1	–	2	–	2	2	–	1	2	2	1.2
Arctagrostis latifolia	–	2	–	–	2	–	–	–	–	2	0.6
Senecio residifolius	–	–	2	–	2	2	–	2	–	–	0.8
Silene acaulis	–	–	–	1	–	–	–	–	–	–	0.1
Pyrola chlorantha	–	–	–	–	–	2	–	–	–	–	0.2
P. grandiflora	2	2	1	2	–	–	2	1	2	2	1.4
Salix myrtillifolia	–	–	–	–	3	–	–	2	–	–	0.5
Moneses uniflora	–	–	–	–	–	2	–	–	–	–	0.2
Cardamine microphylla	–	1	–	–	–	–	–	–	1	–	0.2
Platanthera obtusata	–	–	–	–	2	–	2	–	–	–	0.4
Listera borealis	–	–	–	–	–	–	2	–	–	–	0.2
Senecio atropurpureus ssp. frigidus	–	1	2	–	–	–	–	–	–	1	0.4
Equisetum scirpoides	2	–	–	1	2	–	–	2	–	–	0.7
Epilobium latifolium	–	–	–	–	–	2	–	–	–	–	0.2
Saussurea angustifolia	2	–	2	–	–	–	2	–	–	–	0.6

	Stands										$\overline{X}$
Parrya nudicaulis	-	-	-	-	-	-	-	-	2	-	0.2
Luzula parviflora	1	-	-	1	2	-	1	-	2	1	0.8
Stellaria longipes	-	-	-	-	-	1	-	-	-	-	0.1
Cinclidium stygium	2	1	-	2	2	-	2	2	-	2	1.3
Isopterygium pulchellum	2	-	-	2	1	-	2	-	-	2	0.9
Myurella julacea	2	1	1	-	2	2	1	2	1	1	1.3
Cyrtomnium hymenophylloides	-	2	-	1	-	1	-	1	-	-	0.5
Orthothecium chryseum	-	1	1	-	1	2	1	1	2	-	0.9
Timmia austriaca	2	-	2	-	-	-	-	2	-	-	0.6
Tomenthypnum nitens	2	2	2	2	3	2	1	2	1	2	1.9
Hylocomium splendens	2	2	3	2	1	1	-	2	-	2	1.5
Cetraria nivalis	2	1	-	1	2	-	1	-	-	1	0.8
C. richardsonii	1	2	-	-	-	-	-	1	-	-	0.4
Cladonia alpestris	-	-	2	-	-	-	1	-	-	1	0.4

(d) Main species and their Domin cover values in the ground vegetation of white spruce on uplands with calcareous parent materials and southern aspect; data from Ritchie (1982)

	Stands								$\overline{X}$
Juniperus communis	3	-	4	-	-	4	-	-	1.4
Potentilla fruticosa	3	3	2	-	3	2	-	3	2
Dryas octopetala	2	2	3	2	3	1	3	2	2.25
Arctostaphylos uva-ursi	2	2	2	3	2	2	2	1	2
Rhododendron lapponicum	2	-	2	-	3	2	-	3	1.5
Cassiope tetragona	3	4	2	-	-	2	-	3	1.8
Vaccinium vitis-idaea	2	2	-	2	3	-	2	3	1.8
Ledum groenlandicum	-	-	2	-	-	3	-	-	0.6
Hedysarum alpinum	1	2	-	2	1	2	2	2	1.5
Astragalus umbellatus	-	-	2	-	1	-	1	-	0.5
Melandrium apetalum	-	-	-	1	-	-	-	-	0.1
Carex scirpoidea	2	2	2	3	2	1	2	2	2
Anemone parviflora	2	-	2	-	1	2	2	-	1.1
Carex rupestris	-	2	-	2	2	-	2	-	1
Eritrichium aretioides	-	-	-	-	-	1	-	-	0.1
Bupleurum triradiatum	-	-	2	-	-	-	2	-	0.5
Carex concinna	1	-	-	1	-	2	-	-	0.5
C. glacialis	-	2	-	-	-	-	-	2	0.5
Gentiana propinqua	-	-	-	2	-	-	-	-	0.25
Pedicularis labradorica	-	-	-	-	1	-	1	-	0.25
Hedysarum mackenzii	-	-	2	-	-	2	-	-	0.5
Festuca altaica	-	1	-	-	1	-	1	-	0.4
Tofieldia coccinea	-	-	-	-	-	2	-	2	0.5
Castilleja caudata	-	-	-	2	-	-	-	-	0.25
Androsace chamaejasme	2	-	-	-	2	-	-	-	0.5
Polygonum viviparum	2	-	2	-	-	-	-	-	0.5
Arctostaphylos rubra	-	2	-	-	1	-	-	-	0.4
Lesquerella arctica	-	-	-	2	-	2	-	2	0.75

	Stands								$\overline{X}$
Oxytropis campestris	2	1	–	1	–	1	2	–	0.9
Cnidium cnidiifolium	–	–	–	–	2	–	–	–	0.25
Dodecatheon frigidum	–	–	1	–	–	–	–	–	0.1
Arabis hirsuta var. pynocarpa	–	–	–	–	2	–	–	–	0.25
Dryas integrifolia	–	–	–	–	–	–	1	–	0.1
Campanula aurita	–	–	1	–	1	–	–	–	0.25
Pyrola secunda	–	–	–	–	–	–	2	–	0.25
Pulsatilla patens	1	2	–	–	–	2	–	–	0.6
Geocaulon lividum	–	–	–	–	–	–	2	–	0.25
Woodsia glabella	–	–	–	2	–	–	1	–	0.4
Cypripedium passerium	–	–	–	–	–	–	2	–	0.25
Pedicularis sudetica	–	–	–	–	2	–	–	–	0.25
Zygadenus elegans	1	–	2	2	–	1	–	1	0.9
Senecio lugens	–	2	–	2	–	–	–	2	0.75
Arnica alpina ssp. attenuata	2	–	–	3	–	–	–	–	0.6
Mertensia paniculata	–	–	–	–	–	–	2	–	0.25
Solidago multiradiata	–	–	–	3	–	3	–	–	0.75
Tomenthypnum nitens	2	2	–	3	2	2	2	2	1.9
Rhytidium rugosum	3	–	2	–	–	2	–	–	0.9
Dicranum sp.	2	1	2	–	2	2	–	2	1.4
Distichium capillaceum	–	2	2	–	2	–	2	–	1
Plagiopus oederiana	–	–	–	3	–	–	–	–	0.4
Myurella julacea	–	–	–	–	–	1	2	–	0.4
Cetraria nivalis	2	1	2	–	2	–	–	2	1.1
C. islandica	–	2	1	–	1	–	–	1	0.6
Cladonia mitis	2	–	2	2	–	–	2	2	1.25
Cetraria tilesii	–	–	–	–	2	–	–	–	0.25
Cladonia alpestris	–	–	2	–	–	2	–	–	0.5

(e) Main species and their Domin cover values in the ground vegetation of white spruce / larch stands on uplands with calcareous parent materials in the south Richardson Mountains; data from Ritchie (1982)

	Stands										$\overline{X}$
Alnus crispa	5	–	–	4	–	–	–	–	4	–	1.3
Salix richardsonii	–	4	–	–	4	–	4	–	–	–	1.2
Cassiope tetragona	3	2	4	3	2	3	3	4	2	3	2.9
Dryas octopetala	2	3	2	3	3	3	2	2	3	2	2.5
Vaccinium uliginosum	3	1	–	3	2	2	1	2	3	2	1.9
Salix reticulata	2	–	3	–	2	2	–	2	2	2	1.5
Arctostaphylos rubra	3	3	3	3	2	3	4	3	2	2	2.8
Betula glandulosa	2	–	2	–	–	4	–	–	–	4	1.2
Festuca altaica	2	2	2	1	2	2	2	–	2	2	1.7
Senecio atropurpureus	–	2	–	1	–	–	1	–	1	–	10.5
Tofieldia pusilla	2	1	2	2	–	2	2	2	2	2	1.7
Equisetum scirpoides	3	–	–	–	2	–	–	–	–	–	0.5
Papaver macounii	–	1	2	–	1	2	2	2	–	2	1.2
Lupinus arcticus	2	1	2	2	2	2	2	2	2	2	1.9
Saussurea angustifolia	2	3	2	1	2	2	2	1	2	2	1.9

	Stands										$\overline{X}$
Astragalus umbellatus	–	1	–	1	–	–	2	–	2	1	0.7
Stellaria edwardsii	–	–	1	–	–	1	–	–	–	–	0.2
S. longipes	1	–	–	–	–	–	–	–	–	–	0.1
Ledum groenlandicum	–	2	–	–	2	–	–	2	–	–	0.6
Pedicularis capitata	1	2	2	–	2	–	2	2	–	2	1.3
P. kanei	–	1	–	–	–	–	–	–	–	–	0.1
Carex scirpoidea	2	2	2	2	1	2	–	2	2	2	1.7
C. podocarpa	1	1	2	1	2	–	2	1	–	1	1.1
Pedicularis sudetica	–	–	1	–	–	1	–	–	–	–	0.2
Silene acaulis	2	–	1	–	1	–	–	–	1	–	0.5
Rhododendron lapponicum	–	1	–	2	1	2	–	2	2	2	1.2
Anemone parviflora	2	1	2	–	2	1	2	–	2	2	1.4
Hedysarum alpinum	1	1	2	–	2	–	2	–	2	2	1.2
Salix glauca	2	–	–	–	–	–	–	–	2	–	0.4
Shepherdia canadensis	–	2	–	–	–	–	–	–	–	–	0.2
Corallorhiza trifida	–	–	2	–	–	–	–	–	–	–	0.2
Androsace chamaejasme	–	–	1	–	–	–	–	–	–	–	0.1
Potentilla fruticosa	–	1	–	–	1	–	–	1	–	–	0.3
Ledum decumbens	–	–	–	–	–	1	–	–	–	–	0.1
Zygadenus elegans	–	–	–	–	–	1	–	–	–	–	0.1
Arnica alpina	–	–	–	–	1	–	–	1	–	–	0.2
Polygonum viviparum	–	–	–	–	–	1	–	–	–	–	0.1
Parnassia kotzebuei	–	–	–	–	–	–	–	1	–	–	0.1
Andromeda polifolia	–	–	–	–	–	–	–	–	1	–	0.1
Juniperus communis	–	–	–	1	–	–	–	–	–	–	0.1
Vaccinium vitis-idaea	–	–	–	–	–	–	–	–	1	–	0.1
Oxtropis campestris	–	–	–	–	2	–	–	–	–	1	0.3
Boschniakia rossica	–	–	–	–	–	–	–	–	1	–	0.1
Pedicularis labradorica	–	–	–	–	–	–	–	–	1	–	0.1
Dicranum muehlenbeckii	2	–	1	2	–	2	2	–	2	1	1.2
Tomenthypnum nitens	3	2	3	2	1	3	2	2	2	2	2.2
Hylocomium splendens	2	2	2	2	3	–	2	–	2	3	1.8
Peltigera aphthosa	2	–	2	–	–	2	–	–	2	–	0.8
P. canina	2	–	–	–	–	2	–	–	–	–	0.4
Cetraria nivalis	2	–	2	–	2	–	–	2	2	1	1.1
Cladonia alpestris	–	2	–	2	2	–	2	1	2	2	1.3
Cetraria islandica	2	–	–	1	2	–	2	1	2	1	1.1

(f) Main species and their Domin cover values in the ground vegetation of black spruce open woodlands with shrub and moss; data from Inglis (1975) and Ritchie (1982)

	Stands										$\overline{X}$
Betula glandulosa	4	3	–	3	–	2	–	4	3	3	2.2
Ledum groenlandicum	1	–	3	1	2	1	3	1	1	1	1.2
L. decumbens	2	1	–	2	–	–	–	2	2	1	1
Vaccinium vitis-idaea	3	2	1	2	2	2	1	2	2	3	2
V. uliginosum	3	1	3	1	1	2	2	–	1	–	1.4
Oxycoccus microcarpus	2	–	2	–	3	–	2	–	–	–	0.9

	Stands										$\overline{X}$
Rubus chamaemorus	2	1	–	2	–	–	–	3	2	1	1.1
Carex membranacea	2	–	2	1	–	2	–	1	–	1	0.9
C. lugens	3	–	–	–	4	1	–	2	–	–	1
Equisetum sylvaticum	1	2	–	2	2	–	3	–	2	1	1.3
Calamagrostis canadensis	2	–	1	–	–	3	–	–	–	–	0.6
Arctagrostis latifolia	1	–	–	1	–	–	–	2	–	–	0.4
Eriophorum vaginatum	4	–	–	–	4	–	3	–	–	2	1.3
Carex rostrata	2	–	3	–	–	1	–	–	3	–	0.9
Alnus crispa	–	–	–	5	–	–	1	–	–	–	0.6
Parnassia palustris	–	1	–	–	2	–	–	–	–	–	0.3
Valeriana capitata	–	–	1	–	–	–	1	–	–	–	0.2
Pyrola grandiflora	–	–	–	2	–	2	–	1	–	1	0.6
Empetrum nigrum	2	–	–	2	–	–	–	2	–	2	0.8
Andromeda polifolia	–	2	1	–	2	–	2	–	–	–	0.7
Pedicularis verticillata	–	–	2	–	–	–	–	2	–	–	0.4
P. labradorica	–	–	–	1	–	–	3	–	–	–	0.4
Pinguicula villosa	–	–	1	–	–	–	–	–	–	1	0.2
Salix glauca	–	–	1	4	–	–	–	–	–	–	0.5
Pedicularis lapponica	–	–	–	2	–	3	–	–	1	–	0.6
Arctostaphylos rubra	–	–	–	–	1	–	1	–	1	2	0.5
Equisetum arvense	–	–	–	–	1	1	–	1	1	1	0.5
Saussurea angustifolia	–	–	–	–	2	1	–	–	–	–	0.3
Pyrola secunda	–	–	–	–	–	1	–	1	–	1	0.3
Rosa acicularis	–	–	–	–	–	–	–	1	1	1	0.3
Pinguicula vulgaris	–	–	–	–	–	–	–	–	1	–	0.1
Equisetum scirpoides	–	–	–	–	–	–	–	–	–	1	0.1
Petasites frigidus	–	–	–	–	–	–	–	–	–	1	0.1
Tofieldia pusilla	–	–	–	–	–	–	–	–	–	1	0.1
Sphagnum fuscum	4	3	–	3	–	2	–	3	2	–	1.7
S. rubellum	2	–	2	–	–	–	2	–	–	–	0.6
Aulacomnium turgidum	2	2	2	1	2	3	–	2	3	2	1.9
Polytrichum piliferum	1	3	–	–	2	–	2	–	–	–	0.8
Dicranum acutifolium	1	–	2	–	–	–	–	2	–	1	0.6
Hylocomium splendens	1	2	–	–	–	2	–	–	3	–	0.8
Dicranum elongatum	1	–	4	1	2	3	–	1	–	1	1.3
Sphenobolus minutus	1	1	–	–	1	3	–	–	–	–	0.6
Sphagnum balticum	–	4	–	5	–	3	–	–	–	–	1.2
Tomenthypnum nitens	–	3	–	1	1	3	2	1	2	1	1.4
Ptilidium ciliare	–	1	–	1	1	2	1	–	1	1	0.8
Drepanocladus uncinatus	–	–	2	–	–	2	1	–	–	–	0.5
Sphagnum warnstorfii	–	–	1	1	1	1	1	–	–	–	0.5
Dicranum scoparium	–	–	1	2	–	–	1	–	–	–	0.4
Polytrichum strictum	–	–	–	1	1	–	1	–	–	1	0.4
Polytrichum juniperinum	–	–	–	1	1	–	1	–	–	1	0.4
Dicranum fuscescens	–	–	–	1	–	–	–	–	–	–	0.1
Aulacomnium palustre	–	–	–	–	2	1	1	2	1	1	0.8
Drepanocladus exannulatus	–	–	–	–	1	–	–	1	–	–	0.2
Cetraria nivalis	1	1	–	2	–	–	–	2	–	–	0.6
Cladonia rangiferina	1	2	2	2	2	1	3	–	2	–	1.5

	Stands										$\overline{X}$
C. alpestris	2	1	–	–	2	–	2	1	–	2	1
Cladonia mitis	–	3	–	3	5	–	–	4	3	–	1.8
Peltigera aphthosa	–	1	–	1	–	2	–	4	–	1	0.9
Cladonia alpestris	–	1	–	1	–	1	2	1	–	–	0.6
Hypogymnia physodes	–	–	1	–	1	–	–	1	–	–	0.3
Cetraria sepincola	–	–	1	–	–	1	1	–	1	1	0.5
Cetraria pinastri	–	–	–	1	–	1	1	2	–	–	0.5
Parmeliopsis ambigua	–	–	–	1	–	1	–	1	1	–	0.4
Cetraria cucullata	–	–	–	1	–	1	–	–	–	–	0.2
Cladonia gracilis	–	–	–	–	1	–	–	1	–	–	0.2
Cetraria islandica	–	–	–	–	1	1	2	–	1	1	0.6
Cladonia chlorophaea	–	–	–	–	1	–	1	–	–	–	0.2
Alectoria lanestris	–	–	–	–	–	1	–	–	–	–	0.1

(g) Main species and their Domin cover values in the *Salix phlebophyllum* tundra type based on data in Lambert (1968) and my own records, including Ritchie (1982)

	Stands					$\overline{X}$
Salix phlebophylla	4	4	6	4	4	4.4
Hierochloe alpina	1	2	2	1	1	1.4
Arenaria arctica	1	2	1	1	1	1.2
Oxytropis nigrescens	3	2	–	–	1	1.2
Dryas octopetala	4	1	–	1	1	1.4
Carex podocarpa	–	1	3	1	1	1
Antennaria neoalaskana	2	1	–	1	1	1
Douglasia arctica	1	1	1	–	2	1
Selaginella sibirica	1	1	1	–	1	0.8
Arctostaphylos alpina	–	1	–	1	1	0.6
Artemisia arctica	–	1	–	–	1	0.4
Luzula confusa	–	–	–	1	1	0.4
Smelowskia calycina	–	1	–	–	–	0.2
Gymnomitrion corallioides	1	1	–	2	4	1.6
Polytrichastrum alpinum	–	1	1	–	1	0.6
Polytrichum piliferum	2	2	–	–	–	0.8
P. juniperum	–	–	1	–	5	1.2
Rhacomitrium lanuginosum	–	–	–	2	1	0.6
Cornicularia divergens	2	3	4	2	3	2.8
Alectoria miniscula	2	2	1	1	1	1.4
Parmelia separata	1	2	2	1	1	1.4
P. omphalodes	1	1	1	1	1	1
Sphaerophorus globosus	1	–	1	–	1	0.6
Rhizocarpon geographicum	1	2	3	4	–	2
Haematomma lapponicum	1	2	4	–	2	1.8
Cetraria nivalis	1	1	2	1	1	1.2
C. chrysantha	1	1	–	2	1	1
C. cucullata	1	–	–	1	1	0.6

(h) Main species and their Domin cover values in stands of *Dryas–Carex scirpoidea* tundra on calcareous uplands; data from Ritchie (1982)

	Stands										$\overline{X}$
Dryas octopetala	5	4	4	3	5	4	2	4	4	4	3.9
Arctostaphylos rubra	3	3	2	3	4	3	3	2	2	2	2.7
Cassiope tetragona	-	-	4	-	1	-	3	-	4	-	1.2
Astragalus umbellatus	2	3	1	2	2	2	2	1	3	2	2.0
Carex scirpoides	3	2	3	2	2	3	3	3	2	1	2.4
Oxytropis maydelliana	2	1	1	1	2	2	1	1	-	2	1.3
Salix reticulata	-	2	3	1	2	-	2	3	-	2	1.5
S. glauca	3	2	2	3	-	3	2	2	3	3	2.3
Kobresia myosuroides	2	1	2	2	1	1	2	1	1	2	1.5
Tofieldia pusilla	-	2	-	-	2	-	-	2	2	-	0.8
Pedicularis capitata	-	1	1	2	-	1	2	-	1	1	0.9
P. kanei	1	2	1	1	1	2	1	1	1	1	1.2
Polygonum viviparum	1	1	2	1	1	2	1	2	1	1	1.3
Saxifraga oppositifolia	1	-	-	-	-	-	-	-	-	-	0.1
Carex capillaris	1	1	-	-	1	-	-	-	-	-	0.3
Lupinus arcticus	-	2	-	2	2	-	2	-	2	2	1.2
Papaver macounii	-	-	2	-	-	-	-	2	-	-	0.4
Arctagrostis latifolia	-	2	-	1	-	2	-	-	2	-	0.7
Senecio resedifolius	1	1	1	1	1	1	1	1	1	1	1
Silene acaulis ssp. acaulis	1	-	-	-	1	-	-	-	1	-	0.3
Epilobium latifolium	-	-	-	-	2	-	-	-	-	-	0.2
Saxifraga hieracifolia	-	-	2	-	-	-	-	2	-	-	0.4
Luzula arctica	-	1	-	1	-	-	-	-	-	1	0.3
Parrya nudicaulis ssp. interior	-	-	-	-	-	2	-	-	-	-	0.2
Dryas integrifolia	4	4	5	5	6	5	4	5	4	4	4.6
Hedysarum mackenzii	2	1	1	2	1	1	2	3	1	1	1.5
Anemone drummondii	1	2	1	2	1	1	2	1	2	1	1.4
Tofieldia coccinea	1	1	-	1	2	-	-	1	-	1	0.7
Gentiana propinqua	1	-	-	-	-	-	-	-	-	-	0.1
Oxytropis campestris	1	2	1	-	1	2	1	-	1	1	1
Anemone parviflora	1	-	-	-	-	-	-	-	-	-	0.1
Eritrichium aretioides	1	2	1	-	-	-	1	1	-	1	0.7
Androsace chamaejasme ssp. lehmanniana	-	1	-	-	-	1	-	-	1	-	0.3
Campanula uniflora	-	-	2	-	-	-	-	1	-	-	0.3
Carex misandra	-	-	-	-	1	-	-	-	-	-	0.1
C. petricosa	-	-	-	-	2	-	-	-	-	-	0.2
Antennaria friesiana ssp. compacta	-	-	-	-	1	-	-	-	-	-	0.1
Lesquerella arctica	-	-	-	-	-	1	-	-	-	-	0.1
Rhododendron lapponicum	-	-	-	-	-	-	1	-	-	-	0.1
Hylocomium alaskanum	2	3	2	-	1	2	2	3	2	2	1.9
Cinclidium stygium	-	2	2	2	1	2	-	2	-	2	1.3
Isopterygium pulchellum	1	2	1	1	2	-	1	2	1	2	1.2
Myurella julacea	1	-	2	1	-	-	2	-	1	1	0.8
Cyrtomnium hymenophylloides	-	-	-	1	-	-	2	-	-	-	0.3
Orthothecium chryseum	-	-	1	-	-	2	1	-	-	-	0.4
Timmia austriaca	-	-	-	2	-	-	-	2	-	-	0.4

	Stands										X
Tomenthypnum nitens	2	3	1	–	2	–	2	–	2	2	1.4
Ditrichum flexicaule	–	2	1	2	–	1	–	1	–	1	0.8
Cetraria tilesii	–	–	3	–	1	–	–	3	–	1	0.8
C. cucullata	1	2	1	2	2	3	1	1	1	1	1.5
C. nivalis	2	1	3	4	2	4	1	2	3	4	2.6
Asahinea chrysantha	1	2	1	1	2	1	1	1	2	1	1.3
Cornicularia divergens	1	–	1	2	–	2	1	1	–	1	0.9
Cladonia rangiferina	–	1	–	–	2	–	1	–	–	3	0.7

(i) Main species and their Domin cover values in stands of *Betula-Ledum* tundra; data derived from Lambert (1968), Ritchie (1982), and my own records

	Stands										$\overline{X}$
Betula glandulosa	4	3	3	4	4	1	4	2	3	3	3.1
Arctostaphylos alpina	2	1	3	1	1	2	1	3	3	2	1.9
Ledum decumbens	3	2	2	3	–	1	3	2	2	3	2.1
Salix arctica	2	3	1	3	2	3	2	2	2	3	2.3
S. phlebophylla	1	1	3	1	2	3	1	3	2	3	2.0
Dryas octopetala	1	1	2	–	1	–	2	–	3	1	1.1
Empetrum nigrum	3	–	1	2	1	1	2	3	1	1	1.3
Artemisia arctica	1	–	1	2	1	1	–	1	2	1	1.0
Luzula confusa	1	1	–	1	–	1	1	–	1	1	0.7
Silene acaulis	–	–	3	–	3	–	–	3	2	1	1.2
Rhododendron lapponicum	1	–	–	–	–	1	–	–	1	–	0.3
Loiseleuria procumbens	1	–	1	–	2	2	–	1	–	1	0.8
Oxytropis nigrescens	2	2	1	2	–	1	–	2	–	1	1.1
Potentilla elegans	1	1	3	–	1	1	–	3	1	1	1.2
Pedicularis lanata ssp. *lanata*	1	–	1	1	2	–	2	–	1	1	0.9
Polygonum bistorta ssp. *plumosum*	1	1	2	1	1	2	1	1	1	1	1.2
Hierochloe alpinum	1	1	1	1	–	1	–	1	–	1	0.7
Festuca rubra	–	1	–	1	–	1	–	–	1	–	0.4
Carex microchaeta	1	1	–	1	–	1	–	1	–	1	0.6
C. podocarpa	–	1	1	–	1	–	–	–	1	–	0.4
Draba palanderiana	–	1	–	1	–	–	1	–	–	1	0.4
Eutrema edwardsii	–	–	1	–	–	2	–	–	–	–	0.3
Antennaria friesiana ssp. *compacta*	–	–	2	–	–	–	–	3	–	–	0.5
Minuartia arctica	–	–	1	–	1	2	–	–	–	2	0.6
Papaver lapponicum	–	–	1	–	–	1	–	1	–	–	0.3
Smelowskia calycina	–	–	–	1	–	–	1	–	–	–	0.2
Lagotia glauca ssp. *minor*	–	–	–	–	2	–	–	–	–	–	0.2
Juncus biglumis	1	–	–	–	–	1	–	–	1	–	0.3
Lycopodium selago ssp. *appressum*	–	1	–	2	–	–	–	1	–	–	0.4
Campanula lasiocarpa	–	–	1	–	1	–	–	–	–	–	0.2
Potentilla uniflora	–	–	–	1	–	–	1	–	–	–	0.2
Silene repens	–	–	–	–	1	–	–	–	–	–	0.2
Dryopteris fragrans	–	–	–	–	–	1	–	–	–	–	0.1
Vaccinium vitis-idaea	3	2	1	3	6	3	6	4	4	3	3.5

	Stands										$\bar{X}$
Vaccinium uliginosum	5	1	1	1	5	3	1	–	2	2	2.1
Cassiope tetragona	–	1	–	–	–	–	1	–	–	1	0.3
Poa arctica	–	1	–	–	–	–	1	–	–	1	0.3
Arctagrostis latifolia	–	1	–	–	–	–	2	–	–	1	0.4
Eriophorum vaginatum	–	–	–	–	1	1	1	–	–	–	0.3
Carex lugens	–	–	–	–	–	1	1	–	–	–	0.2
Diapensia lapponica	–	–	1	–	–	–	1	–	2	–	0.4
Saussurea angustifolia	–	–	–	–	–	–	–	–	–	–	
Arenaria arctica	–	–	–	–	–	–	1	–	1	–	0.2
Pedicularis labradorica	–	–	1	1	–	1	–	–	–	–	0.3
Polygonum viviparum	–	–	1	–	1	–	–	–	–	–	0.2
Douglasia arctica	–	–	1	–	–	–	1	–	↲	–	0.2
Anemone narcissiflora	–	1	–	–	–	–	1	–	–	–	0.2
Petasites frigidus	–	–	–	–	–	–	–	–	–	–	
Lupinus arcticus	–	–	–	–	–	–	–	–	–	1	0.1
Antennaria neoalaskana	2	1	–	1	1	–	–	1	–	1	0.7
Selaginella sibirica	1	1	1	–	1	–	–	1	–	1	0.6
Rhacomitrium lanuginosum	1	–	2	1	–	–	–	1	–	1	0.6
Polytrichastrum alpinum	1	–	1	1	–	1	–	–	1	–	0.5
Schistidium gracile	–	–	–	–	–	–	1	–	–	1	0.2
Dicranum groenlandicum	1	–	1	2	–	1	–	1	2	1	0.9
Aulacomnium turgidum	1	1	–	1	2	–	1	1	1	1	0.9
Dicranum elongatum	3	4	2	3	–	4	1	3	2	1	2.3
Polytrichum juniperinum	6	1	1	1	–	1	1	5	–	2	1.8
Sphenolobus minutus	1	1	1	1	–	1	1	–	–	1	0.7
Hylocomium splendens	3	–	–	–	–	–	–	–	–	–	0.3
Dicranum fuscescens	3	–	–	–	2	2	–	1	1	–	0.9
Ptilidium ciliare	1	–	–	3	1	1	–	–	2	1	0.9
Rhytidium rugosum	4	–	–	–	–	1	–	2	–	–	0.7
Sphagnum girgensohnii	–	–	–	–	–	–	–	–	–	–	
Dicranum scoparium	–	–	–	2	–	–	–	–	–	–	0.2
Aulacomnium palustre	–	–	–	–	–	–	2	–	–	–	0.2
Dicranum angustum	1	–	–	2	–	–	–	–	–	–	0.3
Polytrichum piliferum	–	–	–	–	–	–	–	–	–	–	
Gymnomitrion corallioides	–	–	–	–	–	–	–	1	–	–	0.1
Cetraria nivalis	2	1	1	2	2	1	2	1	1	1	1.4
Cladonia mitis	1	1	–	1	1	–	2	1	–	1	0.8
Cladonia alpestris	3	–	1	–	1	–	2	–	1	1	0.9
Cetraria richardsonii	–	–	2	–	1	–	–	1	–	1	0.5
Peltigera canina	–	–	1	–	–	–	1	–	–	–	0.2
Cetraria cucullata	1	6	4	4	1	4	1	3	6	2	3.2
Cetraria islandica	1	1	–	3	1	1	1	1	1	1	1.1
Cladonia rangiferina	1	–	1	5	4	–	1	–	1	1	1.4
Thamnolia vermicularis	–	–	2	–	1	–	1	1	–	1	0.6
Dactylina arctica	1	1	–	–	–	1	1	1	1	1	0.7
Stereocaulon alpinum	–	–	–	–	5	–	4	–	–	–	0.9
Cornicularia divergens	1	1	5	–	–	1	2	3	–	1	1.4
Alectoria ochroleuca	–	–	–	–	–	–	1	2	6	–	0.9
Sphaerophorus globosus	–	1	1	–	–	–	1	1	–	–	0.4

	Stands										$\overline{X}$
Ochrolechia frigida	-	-	3	-	-	-	3	1	1	-	0.8
Rhizocarpon geographicum	-	-	2	-	-	-	1	1	1	-	0.5
Cladonia gracilis	1	-	-	1	-	-	1	1	-	-	0.4
Cetraria nigricascens	-	-	1	-	-	-	1	-	-	-	0.2
Peltigera aphthosa	-	-	-	-	-	-	1	-	-	-	0.1

(j) Main species and their Domin cover values in stands of the *Betula glandulosa–Eriophorum* tundra type; data from Lambert (1968) and my own records

	Stands														$\overline{X}$
Eriophorum vaginatum	7	6	6	8	8	5	8	6	7	8	6	6	8	7	6.9
Betula glandulosa	5	3	3	2	3	1	3	3	5	5	4	-	4	3	3.1
Ledum decumbens	4	3	1	2	4	3	5	4	5	5	3	-	+	1	2.9
Vaccinium vitis-idaea	5	2	2	1	5	1	5	3	4	5	3	-	1	3	2.9
Empetrum hermaphroditum	3	2	1	1	4	1	-	2	-	3	1	-	2	2	1.6
Vaccinium uliginosum	2	2	2	2	1	1	1	1	2	3	3	2	2	-	1.7
Carex lugens	-	4	4	-	5	1	2	1	-	4	2	1	4	5	2.4
Rubus chamaemorus	-	1	2	1	3	3	4	3	2	2	2	-	-	-	1.6
Pedicularis lapponica	-	2	1	1	1	1	2	1	-	2	3	-	-	-	1
Salix pulchra	-	1	5	-	-	2	-	-	-	3	1	1	4	4	1.5
Polygonum bistorta	-	1	-	-	2	-	-	2	-	2	-	2	2	1	0.9
Arctostaphylos alpina	-	3	-	-	4	-	-	2	-	-	-	-	-	2	0.8
Salix reticulata	-	-	2	-	-	-	-	-	-	-	-	-	-	2	0.3
Andromeda polifolia	1	-	-	2	-	-	-	-	-	-	2	-	-	-	0.4
Pinguicula villosa	-	-	-	2	1	-	-	-	-	1	2	-	-	-	0.4
Luzula nivalis	1	-	-	-	1	-	-	-	-	-	-	+	1	-	0.2
Arctagrostis latifolia	-	-	-	-	-	-	-	2	-	-	-	+	3	-	0.4
Petasites frigidus	-	-	-	-	-	-	-	2	-	-	-	2	2	-	0.4
Carex rotundata	-	-	-	1	-	-	-	-	-	-	3	1	-	-	0.4
Cassiope tetragona	-	-	-	-	-	-	-	-	-	-	-	-	+	-	
Dryas octopetala	-	-	-	-	-	-	-	-	-	-	-	2	1	1	0.3
Salix glauca	-	-	-	-	-	-	-	-	-	-	-	2	2	1	0.4
Lagotis glauca	-	-	-	-	-	-	-	-	-	-	-	2	2	-	0.3
Poa arctica	-	-	-	-	-	-	-	-	-	-	-	+	1	+	0.07
Stellaria ciliatosepala	-	-	-	-	-	-	-	-	-	-	-	1	+	1	0.1
Pyrola grandiflora	-	-	-	-	-	-	-	-	-	-	-	-	2	1	0.2
Hierochloe alpina	1	-	-	-	-	-	-	1	-	-	-	1	-	-	0.2
Pedicularis lanata	-	-	-	-	-	-	-	1	-	-	-	+	+	-	0.07
Salix phlebophylla	-	-	-	-	-	-	-	2	-	-	-	1	-	-	0.2
Oxytropis maydelliana	-	-	1	-	-	-	-	-	-	-	-	+	-	1	0.1
Saxifraga punctata	-	-	-	-	-	-	-	-	-	-	-	1	1	-	0.1
Luzula confusa	-	-	-	-	-	-	1	-	-	-	-	1	-	-	0.1
Pedicularis labradorica	-	-	+	-	-	-	-	-	-	-	-	-	-	+	
Juncus biglumis	-	-	-	-	-	-	-	-	-	-	-	+	-	-	
Senecio atropurpureus	-	-	-	-	-	-	-	-	-	-	-	1	-	1	0.1
Saxifraga hieracifolia	-	-	-	-	-	-	-	-	-	-	-	1	+	1	0.1
Polygonum viviparum	-	-	-	-	-	-	-	-	-	-	-	1	1	-	0.1
Draba nivalis	-	-	-	-	-	-	-	-	-	-	-	+	+	-	

	Stands														$\bar{X}$
Eutrema edwardsii	-	-	-	-	-	-	-	-	-	-	-	1	1	-	0.1
Aulacomnium turgidum	1	3	2	1	2	1	2	3	3	2	2	4	5	5	2.6
Sphagnum lenense	5	4	5	5	7	7	7	8	1	9	6	-	-	-	4.6
Sphagnum rubellum	2	4	2	3	6	-	-	-	-	3	3	-	1	2	1.9
Dicranum angustum	-	2	-	3	-	-	5	1	4	1	4	-	+	2	1.6
Sphenolobus minutus	1	-	1	-	1	+	2	-	-	+	4	1	+	1	0.8
Polytrichum juniperium	3	1	2	1	2	+	1	-	1	-	-	+	-	+	0.8
Sphagnum recurvum	-	4	5	-	-	6	+	5	-	-	2	-	-	3	1.8
Hylocomium splendens	1	2	-	-	-	1	-	-	-	-	-	5	5	4	1.3
Dicranum elongatum	3	-	1	-	-	1	-	-	-	2	2	1	1	2	0.9
Sphagnum girgensohnii	-	4	-	-	-	-	-	-	2	-	4	-	-	-	0.7
Dicranum scoparium	3	2	+	-	-	1	-	-	5	-	-	-	-	-	0.8
Sphagnum balticum	2	-	-	+	-	3	-	2	-	1	1	-	-	-	0.6
Dicranum fuscescens	3	-	-	-	-	-	1	-	-	-	-	-	3	-	0.5
Ptilidium ciliare	-	-	-	-	2	-	-	-	-	-	-	2	3	1	0.6
Calypogeia trichomanis	-	-	-	-	1	-	3	-	-	1	+	-	-	-	0.4
Rhacomitrium lanuginosum	1	2	-	-	-	-	-	-	-	-	-	-	-	1	0.3
Dicranum groenlandicum	-	-	-	3	4	-	-	-	-	1	-	-	-	-	0.6
Tomenthypnum nitens	-	-	-	-	-	-	-	-	-	-	-	5	3	4	0.9
Aulacomnium palustre	-	-	5	-	-	-	-	-	-	-	-	-	-	+	0.4
Rhytidium rugosum	-	2	-	-	-	+	-	-	-	-	-	-	-	-	0.1
Cetraria cucullata	2	+	+	2	3	-	6	1	3	1	2	+	+	1	1.5
Cladonia mitis	+	+	-	1	1	-	4	-	2	1	1	+	+	1	0.8
Cetraria islandica	-	+	+	1	+	-	-	1	+	1	1	+	+	1	0.4
Cladonia rangiferina	3	-	-	1	3	-	4	+	2	-	-	-	-	-	0.9
Peltigera aphthosa	1	2	1	-	-	-	1	1	-	-	-	2	1	3	0.9
Cladonia amaurocraea	-	+	+	-	2	-	4	-	-	1	1	1	1	1	0.8
Dactylina arctica	2	1	-	+	2	-	1	1	+	-	1	-	+	-	0.6
Peltigera scabrosa	1	1	1	-	+	-	1	1	-	-	-	-	1	-	0.4
Thamnolia vermicularis	2	-	-	-	+	1	1	-	-	1	-	-	-	1	0.4
Cetraria nivalis	1	-	-	-	1	-	-	-	1	+	1	-	-	-	0.3
Peltigera polydactyla	-	-	-	-	-	1	1	-	+	-	+	-	+	-	0.1
Peltigera canina	-	-	1	-	-	-	-	1	-	-	-	1	1	1	0.4
Stereocaulon alpinum	-	-	-	-	-	1	-	-	-	-	1	-	-	1	0.2
Cladonia gracilis	-	-	-	-	1	-	-	-	-	-	1	1	-	+	0.2
Alectoria ochroleuca	-	+	-	-	1	-	-	-	-	1	-	-	-	-	0.1
Stereocaulon rivulorum	-	+	-	-	-	-	+	-	+	-	-	-	-	-	
Cladonia furcata	-	-	+	-	-	-	-	-	+	-	+	-	-	-	
Cetraria richardsonii	-	-	-	-	-	-	-	-	1	-	+	+	-	-	0.07
Cladonia chlorophaea	+	-	-	-	-	-	-	-	-	-	1	-	-	+	0.07
Nephroma expallidum	-	-	-	-	-	-	-	-	-	-	-	+	-	-	
Alectoria nitidula	-	+	-	-	1	-	-	-	-	-	-	-	+	-	0.07
Alectoria nigricans	1	-	-	-	-	-	-	-	+	-	-	-	-	-	0.07

(k) Main species and their Domin cover values in stands of *Eriophorum vaginatum* tundra; data from Lambert (1968) and my own records

	Stands								$\overline{X}$
Eriophorum vaginatum	7	6	6	8	5	6	7	8	6.6
Betula glandulosa	5	3	3	2	3	3	5	4	3.5
Ledum decumbens	4	3	1	2	4	3	4	3	3
Vaccinium vitis-idaea	5	2	2	1	1	5	3	4	2.9
Empetrum hermaphroditum	3	2	1	1	4	1	2	1	1.9
Vaccinium uliginosum	2	2	2	1	1	1	2	3	1.8
Carex lugens	–	4	4	–	5	1	1	1	2
Rubus chamaemorus	–	1	2	1	3	3	1	1	1.5
Pedicularis lapponica	–	2	1	1	1	1	2	1	1.1
Salix pulchra	–	1	5	–	–	2	–	–	1
Polygonum bistorta	–	1	–	–	2	–	–	2	0.75
Arctostaphylos alpina	–	3	–	–	4	–	–	2	1.1
Salix reticulata	–	2	–	1	–	2	1	1	0.9
Andromeda polifolia	1	–	–	2	–	–	1	1	0.6
Pinguicula villosa	–	–	–	2	1	–	–	1	0.5
Aulacomnium turgidum	1	3	2	1	1	2	3	1	1.75
Sphagnum lenense	5	4	5	5	7	7	8	6	5.9
S. rubellum	2	4	2	3	2	–	–	–	1.6
Dicranum angustum	–	2	–	3	–	–	5	1	1.4
Sphenolobus minutus	1	–	1	–	1	–	2	–	0.6
Polytrichum juniperinum	3	1	2	–	2	–	1	–	1.1
Sphagnum recurvum	–	4	5	–	4	–	5	–	2.25
Hylocomium alaskanum	1	1	2	–	1	–	1	2	1
Sphagnum girgensohnii	–	4	–	–	–	–	2	–	0.75
Dicranum scoparium	3	2	–	–	1	–	–	–	0.75
D. fuscescens	3	–	–	–	–	1	–	–	0.5
Ptilidium ciliare	–	–	–	2	–	–	–	–	0.25
Cetraria cucullata	2	–	1	1	–	1	3	–	1
C. islandica	–	–	1	1	–	1	–	1	0.5
Cladonia mitis	–	–	1	1	–	1	1	1	0.6
Peltigera aphthosa	1	2	1	1	2	1	1	1	1.25

(l) Main species and their Domin cover values in stands of *Carex aquatilis* wet meadow communities; data abstracted from Lambert (1968)

	Stands														$\overline{X}$
Carex aquatilis	7	7	8	3	7	5	6	5	7	8	8	6	8	9	6.7
Salix arbutifolia	1	–	1	1	4	4	–	3	–	–	–	5	8	5	2.3
Potentilla palustris	4	–	–	5	5	3	4	–	–	3	3	–	–	–	1.9
Eriophorum angustifolium	–	–	–	–	2	1	–	–	6	6	2	–	–	–	1.2
Carex chordorrhiza	–	6	–	7	–	4	–	–	–	–	–	–	–	–	1.2
Eriophorum scheuchzeri	–	1	1	1	5	–	–	1	–	–	–	–	–	–	0.6
Carex rariflora	–	–	–	–	1	–	–	–	–	–	–	–	5	3	0.6

	Stands														$\overline{X}$
Pedicularis sudetica	–	–	–	–	1	1	–	2	–	–	–	+	–	–	0.3
Betula glandulosa	–	–	–	–	–	–	–	–	–	–	–	2	2	3	0.5
Salix pulchra	–	–	–	–	–	–	–	–	–	–	–	1	2	4	0.5
Andromeda polifolia	–	–	–	–	–	–	–	–	–	–	–	1	1	4	0.4
Vaccinium uliginosum	–	–	–	–	–	–	–	–	–	–	–	–	1	4	0.4
Ranunculus pallasii	–	–	–	–	1	+	1	–	–	–	–	–	–	–	0.1
Calamagrostis canadensis	–	–	–	–	–	–	–	–	1	1	–	–	–	–	0.1
Pedicularis lapponica	–	–	–	–	–	–	–	–	–	–	–	+	1	–	0.07
Ledum decumbens	–	–	–	–	–	–	–	–	–	–	–	+	–	1	0.07
Sphagnum squarrosum	–	–	–	4	5	6	–	8	–	–	1	–	6	–	2.1
Polytrichum commune	4	–	–	–	–	–	–	1	–	1	1	1	3	3	1
Sphagnum recurvum	–	–	–	–	–	–	–	–	–	–	–	6	8	7	1.5
Sphagnum platyphyllum	–	–	–	3	4	2	4	–	–	–	–	–	–	–	0.9
Aulacomnium palustre	–	–	–	–	1	2	1	1	1	1	1	2	4	1	1.1
Aulacomnium turgidum	–	–	–	–	–	–	–	–	–	–	–	1	4	3	0.6
Sphagnum rubellum	–	–	–	1	–	–	–	1	–	–	–	3	–	–	0.4
Sphagnum teres	–	–	–	–	–	5	–	5	–	–	–	–	–	–	0.7
Sphagnum contortum	–	2	6	–	–	–	–	–	–	–	–	–	–	–	0.6
Drepanocladus aduncus	–	–	–	–	–	–	–	–	–	2	6	–	–	–	0.6
Sphagnum lenense	–	–	–	–	–	–	–	–	–	–	–	3	–	5	0.6
Calliergon cordifolium	–	–	–	–	–	–	–	4	–	2	–	–	–	–	0.4
Mnium punctatum	–	–	–	–	–	–	–	2	–	1	–	–	–	–	0.2
Cinclidium subrotundum	–	–	–	1	1	–	–	–	–	–	–	–	–	–	0.1

APPENDIX 3

Field and laboratory methods

Readers unfamiliar with the practice of plant palaeoecology might find useful the following description of some aspects of our field, laboratory, and analytical methods, particularly those that differ from those of other workers, as described in Faegri and Iversen (1975), Birks and Birks (1980), Berglund (1979a), and Wright (1980).

The area is large and difficult of access, and with the exception of the land immediately adjacent to the Dempster Highway (Figure 12) all travel on coring trips must be by aeroplane or helicopter. When feasible we try to examine an area during the summer, and to locate and sound lakes and ponds that show promise. An inflatable dinghy with a small (2 hp) motor and a continuously operating recording fathometer (Raytheon, model D-17) are used to make controlled traverses of lakes. This low-frequency sounder provides a continuous graph of the profile of the soft bottom of lakes, and in some conditions can yield both soft and hard bottom traces. In this way the lake morphometry and sediment configuration are established, and a decision can be taken about the potential value of the site for a coring operation. In general we prefer small (5–15 ha), reasonably deep (>5 m) ponds with no significant permanent inlet stream, for the obvious reasons discussed by Jacobson and Bradshaw (1981). We avoid lakes with any evidence on the shores of past erosion, the extreme example of which is a thermokarst lake.

Summer coring from a raft or from a pair of attached dinghies or canoes has been completed satisfactorily, but we prefer to core lake sediments in late winter from the ice surface. Frozen lakes provide the stability and freedom of movement to ensure a reliable operation, and lake size is no limitation on access by helicopter in winter. Also, the period from approximately 20 April to 10 May is ideal with respect to weather and day length. The maximum values for sunshine hours for anywhere in Canada during April and May occur in our area (~350 hr per month), day length is approximately 18 hours, and the ambient midday temperature is usually between 0° and –10°C.

A two-man crew and all equipment can be carried in a small helicopter (e.g. Bell 206). Our routine procedure is as follows. (1) The ice is drilled manually with a 6 inch Finnish-type auger; ice thickness varies with latitude and yearly weather conditions, but lakes in the wooded zone have about 1.5 m (range 0.8–2 m) and those in the arctic zone 2 m (range 1.5–2.5 m). (2) The uppermost 50–70 cm of sediment is recovered either with a dry-ice freezing sampler (Swain 1973) or in a clear plastic tube, 5 cm in diameter and 1 m long, attached to a Livingstone stationary piston sampler (Livingstone 1955b), modified according to Wright (1968). (3) Sectional casing is then lowered into the hole and allowed to penetrate the sediment to a depth ~25 cm less than the lower level of the first sample recovery. The casing is secured to the ice surface by a collar. It is 9-cm-diameter standard plastic water pipe, in 10-foot lengths, and we have available miscellaneous shorter lengths to enable matching with the particular water depth of any lake. The advantages are its lightness, robustness, and ease of coupling by coarse-threaded outside sleeves. It can be used without difficulty in lakes up to a depth of 20 m. Deeper lakes must be sampled by free-fall gravity samplers. (4) The sediment is recovered with stainless steel tubes 5 cm (or 10 cm) in diameter and 1 m (or 2 m) long, using a Livingstone stationary piston corer of the square-rod type; standard, commercially available 5-foot extensions of magnesium-aluminum alloy (referred to as XRT rods) are used. They are light, strong, and coarse-threaded. We find that manual coring with a two-person crew is rarely adequate to penetrate the stiff late-glacial sediments of high-latitude lakes, so we employ a chain jack (Nilcon Geoteknik) to provide the steady continuous downward drive required for successful coring and the upward pull to recover the sample. Each end of the jack must be securely anchored to the ice by a chain looped through a v-shaped hole drilled in the ice to a depth of about 50 cm. The hoist is compact, light, and has ample capacity for any coring operation of this type. (5) Cores are extruded onto a clean flat surface, wrapped in plastic and aluminum foil, and placed in wooden-core boxes. They are shipped to the laboratory quickly, avoiding any risk of freezing, which greatly distorts the stratigraphy.

Peat deposits also yield useful continuous records for palaeoecology, and there are special requirements in permafrost regions to ensure successful recovery. Peats are perennially frozen at depth in our area, and we use a modified Sipre peat corer (Hughes and Terasmae 1963). This consists of a hollow steel drill piece with two tempered steel cutting teeth; the drill bit dimensions are 7.5 cm inside diameter and 35 cm length. The bit is rotated at speed by a two-stroke gasoline motor. The chief limitation of the method is that the bit will not function in a wet hole, so low-lying bogs with surface water cannot be sampled in summer. Otherwise the unit is efficient, and three persons can sample in duplicate a frozen peat deposit 2–3 m deep in 4–6 hours. The sampler provides continuous, uncontaminated samples of adequate size (7.5 m diameter) for all pollen, macrofossil, and 14-C analyses. The complete unit with two or three persons can be transported in a small helicopter.

Unless undisturbed complete cores are required (e.g. for x-ray or palaeomagnetism analysis), the cores are unwrapped in the laboratory and a preliminary sediment description is completed using standard methods (Berglund 1979a, Chapter 8). Samples of fresh sediment are taken for pollen analysis usually at 2-cm intervals, with a fixed-volume (either 1 ml or 0.5 ml) cylindrical chamber (Birks 1973). Exotic pollen in known concentration is added to provide the basis for later calculation of pollen concentration and influx (methods are given in Berglund 1979a; Birks and Birks 1980; and Maher 1981). Samples are concentrated by removing the fine silts and clays using a dispersant-and-sieving method developed by Cwynar et al. (1979). Various standard numerical methods for zoning pollen diagrams, comparing two sites, comparing modern and sub-fossil data, and applying pollen size statistics are used routinely in our laboratory (methods described in Yarranton and Ritchie 1972; Ritchie and Yarranton 1978; Birks and Birks 1980; and Birks 1979). Samples are processed for plant macrofossil analysis by the standard methods (Birks 1980).

One of the acute and familiar problems of pollen analysis in tundra and subarctic woodland regions is that a few plants are greatly overrepresented quantitatively by their pollen in the sediment (e.g. *Betula*, *Alnus*), some important elements in the vegetation are greatly underrepresented (*Larix*, *Ericales*, *Salix*, legumes), and a significant element of the plant cover, such as many species of Rosaceae, Caryophyllaceae, and other families, is totally unrepresented in many lake sediments. In addition, large fractions of these northern landscapes are, and probably were in the past, covered by lichens and mosses – the former never occur in the sediments in any form, and while spores of the latter do they rarely can be used quantitatively because of low, erratic, and poorly understood spore production. In an attempt to grapple with some of the interpretive difficulties that result from these problems, we use available quantitative data on plant cover for some of our area along with modern pollen spectra to generate correction factors (R-values) developed by Davis (1963) to apply to the fossil record. The R-value of a taxon is an estimate of the ratio of the pollen frequency (%) in sediment samples to the percentage cover or equivalent measure of the taxon in the adjacent landscape. Usually the relative R-values (R_{rel}) are then calculated, based on an R-value of 1 for a selected taxon, so that comparisons can be made between values for taxa from different areas. This procedure has its limitations, fully discussed in Birks and Birks (1980) and Prentice (1982), but it has some value in vegetation reconstruction if used conservatively.

APPENDIX 4

A listing of the radiocarbon age determinations of samples from sites referred to in Chapter 5

The ages have been corrected for 13-C fractionation

Site	Location	Sample level (cm)	14-C age	Lab. number
Tuktoyaktuk-5	69°03′N, 133°27′w	62 – 78	3,630 ± 140	GSC-1338
		125 –132	5,440 ± 140	GSC-1269
		175 –185	8,690 ± 180	GSC-1354
		195 –202	11,500 ± 220	GSC-1237
		240 –260	12,900 ± 170	GSC-1321
Sleet Lake	69°17′N, 133°35′w	60 – 68	1,970 ± 60	GSC-3330
		113 –120	5,270 ± 80	GSC-3323
		193 –200	6,210 ± 60	GSC-3311
		302.5–310	10,400 ± 110	GSC-3307
		382.5–390	12,500 ± 110	GSC-3302
Hendrickson Island	69°32′N, 133°35′w	45 – 50	3,160 ± 60	GSC-1905
		85 – 90	3,180 ± 60	GSC-1970
		125 –130	6,810 ± 80	GSC-1960
		175 –180	9,350 ± 80	GSC-1896-2
		175 –180 (shells)	9,010 ± 100	GSC-1896
Eskimo Lakes	69°25′N, 131°40′w	15 – 20	2,920 ± 130	GSC-1669
		70 – 75	4,530 ± 140	GSC-1724
		80 – 85	6,770 ± 140	GSC-1737
		100 –105	9,500 ± 170	GSC-1717
		185 –190	9,690 ± 250	GSC-1671
Richards Island	69°26′N, 134°30′w	381	9,390 ± 150	GSC-1031
M-Lake	68°16′N, 133°28′w	100 –105	2,860 ± 60	GSC-2087
		122 –128	6,410 ± 90	GSC-2221
		165 –170	8,590 ± 80	GSC-2187
		214 –219	10,000 ± 110	GSC-2172
		229 –235	11,300 ± 90	GSC-2075

Twin Tamarack	68°18′N, 133°25′w	2 – 6	440 ± 60	GSC-3394
		67 – 73	3,640 ± 90	GSC-3377
		119 – 126	5,830 ± 90	GSC-3384
		163 – 172	7,810 ± 100	GSC-3347
		296 – 304	11,600 ± 140	GSC-3346
		320 – 330	13,100 ± 150	GSC-3387
Sweet Little Lake	67°38′N, 132°00′w	0 – 5	1,000 ± 120	GSC-3445
		123 – 127	3,470 ± 100	GSC-3443
		239 – 245	5,740 ± 80	GSC-3439
		305 – 310	8,680 ± 80	GSC-3436
		344 – 351	9,720 ± 100	GSC-3430
		375 – 380	9,550 ± 100	GSC-3419
Hanging Lake	68°23′N, 138°23′w	13 – 18	1,870 ± 180	GSC-2807
		23 – 28	2,930 ± 130	GSC-2744
		28 – 33	5,100 ± 180	GSC-2518
		53 – 58	4,360 ± 150	GSC-2517
		78 – 83	5,640 ± 140	GSC-2515
		103 – 108	7,870 ± 280	GSC-2399
		115.5 – 120.5	10,200 ± 190	GSC-2749
		128 – 133	11,500 ± 250	GSC-2514
		140.5 – 145.5	9,220 ± 140	GSC-2826
		153 – 158	10,500 ± 200	GSC-2503
		178 – 183	10,900 ± 210	GSC-2502
		203 – 208	11,700 ± 150	GSC-2489
		228 – 233	12,600 ± 270	GSC-2500
		253 – 258	12,400 ± 210	GSC-2491
		263 – 268	12,800 ± 320	GSC-2846
		273 – 278	15,800 ± 450	GSC-2388
		278 – 283	17,700 ± 380	GSC-2868
		293 – 298	16,800 ± 320	GSC-2830
		303 – 308	20,200 ± 510	GSC-2482
		318 – 323	20,200 ± 660	GSC-2790
		323 – 336.5	24,900 ± 600	GSC-2710
Lateral Pond	65°57′N, 135°31′w	109 – 112	6,800 ± 80	GSC-2854
		137 – 141	7,510 ± 170	GSC-2852
		188 – 192	12,100 ± 130	GSC-2808
		225 – 230	14,800 ± 260	GSC-2785
		270 – 275	15,200 ± 230	GSC-2758

PLATES

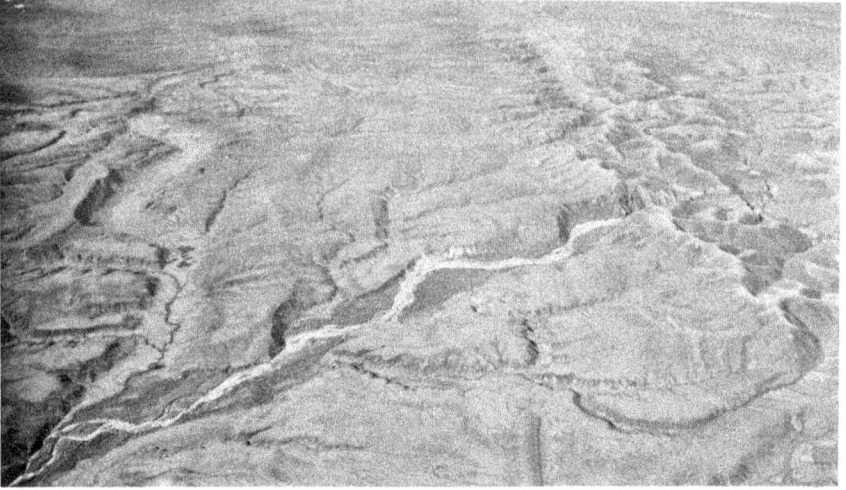

1. An oblique air photograph showing the North Yukon landscapes that lie north of 66° N, between 135°30′ and 136°30′ W. The South Richardson Mts stretch northwards at the right; Canyon Creek flows eastwards from the lower left corner; the abandoned glacial spillway that carried the waters of the Peel River northwards into the Bell, Old Crow, and Bluefish proglacial lake basins during the maximum of Wisconsin glaciation is seen clearly at the left of the photograph. This valley is occupied today by two large lakes, but in full-glacial times it carried the diverted Peel River water into the headwaters of the Eagle River, just visible to the north of the main spillway. Published by permission of the Department of Energy, Mines and Resources (Photo T4-84R)

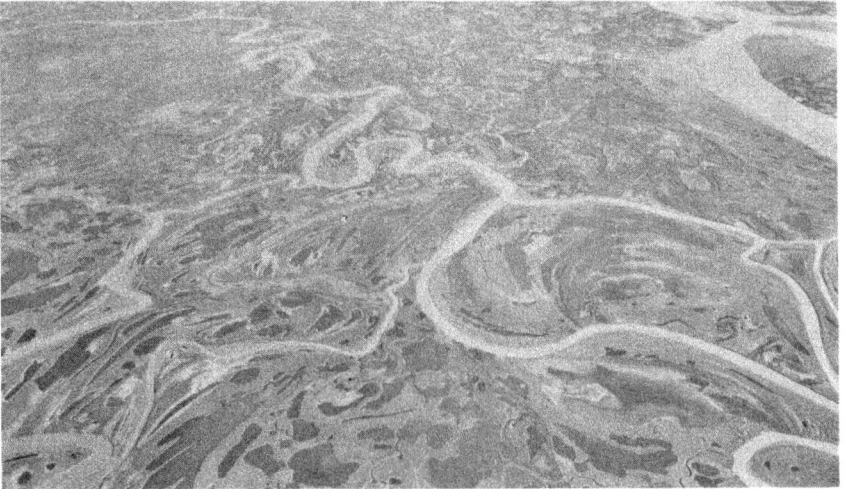

3. An oblique air photograph of the northwest extremity of the Old Crow Flats, looking north towards the British Mts. The lower edge is at about 68°10′ N and the lower right corner is at 140°30′ W, near the Alaska boundary. The Old Crow River bisects the photograph, flowing from left to right, and its confluence with Thomas Creek, which flows south from the British Mts, can be seen near the mid-point of the photograph. The pattern of large, shallow, rectangular, and oriented thermokarst lakes is visible. Published by permission of the Department of Energy, Mines and Resources, Ottawa (Photo T14L-144)

2 (*bottom, facing*). An oblique air photograph of a portion of the Mackenzie Delta, between 134°30′ and 135° W; the lower margin is at roughly 68° N. The Middle Channel of the Mackenzie River is visible on the right and the Aklavik Channel flows north diagonally across the photograph. The 'intricate anastomosing network of channels and interconnecting lakes' with the 'predominance of characteristic channel types' (Mackay 1963, p 105) is obvious. Published by permission of the Department of Energy, Mines and Resources, Ottawa (Photo T3-5R)

4. The Ibyuk pingo, 5 km south of Tuktoyaktuk, showing the typical wide, eroding tension cracks on the summit of an old pingo that is no longer growing. (Mackay 1976)

5. A pattern of irregular stone circles in the upland tundra near the Barn Mountains

6. An aerial view of a field of high-centred ice-wedge polygons near Sitidgi Lake. The centres are covered by a *Cladonia*-heath tundra community.

7. Ice-wedge cracking is visible at the intersection of the wedge ditches in this area of high-centred polygons.

8. Two excavated frost hummocks, the upper in the boreal woodland zone near Inuvik, the lower in the shrub tundra zone near Tuktoyaktuk. Both illustrate the underlying domed permafrost table with upthrusting, fine-grained clays capped by a fibrous humus.

9. Large areas of the forested uplands of the Lower Mackenzie River and adjacent North Yukon are occupied by mixed woodlands of the type shown here. This 100-year-old stand shows the invasion of the mature white birch phase by white spruce. This particular stand is more open than most, and was so chosen for ease of photographic illustration.

12. An example in the upper Eagle River valley of black spruce lichen woodland, here developed over a coarse gravel soil on a terrace feature

10 (*top, facing*). White spruce woodland on an upland site at Dolomite Lake near Inuvik. Openings in the woodland along scarps, as in the foreground, often have local communities of *Juniperus communis*, *Pulsatilla multifida*, and *Shepherdia canadensis*.

11 (*bottom, facing*). White spruce forest on alluvium, Mackenzie Delta near Inuvik

15. Treeline larch woodlands on steep limestone surfaces are locally common in the South Richardson Mts, here at approximately 66° N, 135°42′ W, east of the Doll Creek valley. The sites have a northwest aspect, snow lies deep late into May, and the ground vegetation is dominated by *Cassiope tetragona*, *Dryas octopetala*, and *Tomenthypnum nitens*. White spruce occurs in these communities but on slopes of this particular aspect and bedrock type larch is the chief montane treeline species.

13 (*top, facing*). An upland site in North Yukon near the Eagle River, at 66°30′ N, showing the early willow, alder, and white birch sapling stage of vegetation recovery, approximately 10 years after fire destroyed a white spruce stand

14 (*bottom, facing*). An upland site in North Yukon on the Eagle Plain, 66°30′ N, showing a 25-year-old open white birch community with white spruce regeneration, roughly 30 years after a fire

16. In unglaciated terrain sharp contacts between different rock types can determine abrupt transitions in plant community type. Here, in the South Richardson Mts at about 66°20′ N near the valley of Doll Creek, the contact between a shale-sandstone rock type (more distant) and a limestone type (foreground) produces a spectacular, sharp boundary between two tundra types. The light-coloured tundra in the right foreground is typical *Dryas–Carex scirpoidea* (type (b)); beyond the sharp line is typical *Betula glandulosa–Ledum decumbens* shrub tundra (type (d)); the distant ridge of shale sandstone is covered by *Salix phlebophylla* tundra of type (a).

17. An example of shrubby tussock tundra (*Betula glandulosa–Eriophorum*, type (e) tundra) on a gently sloping surface near the Barn Mts

18. Tussock cotton-grass tundra in a poorly drained lowland site on the North Yukon coastal lowlands, near the Babbage River (type (f))

19. A typical expanse of *Betula glandulosa–Ledum decumbens* tundra (type (d)) on the rolling uplands of the Tuktoyaktuk Peninsula, 12 km north of Parsons Lake

20. Local, apparently clonal patches of *Populus balsamifera* occur several kilometres beyond the limit of continuous tree growth. Here a patch grows on the shoreline of a lake abutting on an esker, at 69°02′ N, 132°27′ W, near Eskimo Lakes.

21. The Old Crow River near Schaeffer Creek, roughly 30 km upstream from Old Crow Village. This riverbank section through the Old Crow Basin sediments is Locality 12, described in detail in Jopling et al. (1981), from which many palaeontological and supposedly archaeological remains have been recovered. It is also one of the sample sites reported in the pollen studies of Lichti-Federovich (1973).

24. Long, roughly east-west trending ridges in the South Richardson Mts show striking vegetation differences between north- and south-facing slopes. Here, on limestone bedrock, the former has a *Dryas-Carex* herb tundra while the latter has an open spruce woodland; these communities are described on pages 52 and 59.

26. Three limestone bedrock ridges intersect to form a rough triangle at this site in the South Richardson Mts, producing slopes of strikingly different vegetation, apparently controlled by differences in the net radiation of each. The opposite slope is south-facing and has a continuous spruce woodland; the foreground slope is northwest-facing with a rich herb tundra; the slope on the left is northeast-facing but is also shaded from the low-angled sun by the ridge behind the camera; it has a fruticose lichen cover with sparse herbs.

25 (*bottom, facing*). A clone of *Picea glauca* formed by layering on the flanks of an esker near Old Man Lake, at 69°08′ N, 132°27′ W.

27. The annual melting of the frozen, supersaturated Pleistocene sediments of the Yukon Coastal Lowlands results in a steady retreat of the coastal scarp, shown here near Stokes Point at 69°25′ N, 139° W.

28. White spruce stump macrofossil on the Tuktoyaktuk Peninsula, at 69°07′ N, 131°33′ W, radiocarbon age 4,940 ± 140 (GSC-1265)

References

Ager, T. 1975. *Late Quaternary Environmental History of the Tanana Valley, Alaska.* Ohio State University, Institute of Polar Studies, Report 54

Ager, T.A. 1982. Vegetational history of western Alaska during the Wisconsin glacial interval and the Holocene. In Hopkins et al. 1982: 75–93

Aleksandrova, V.D. 1980. *The Arctic and Antarctic and Their Division into Geobotanical Areas.* Cambridge University Press, Cambridge

Anderson, J.P. 1959. *Flora of Alaska and Adjacent Parts of Canada.* Iowa State University Press, Ames, Iowa (2nd ed.)

Anon. 1974. *Vegetation Types of the Mackenzie Corridor.* Environmental-Social Northern Pipelines Committee, Report 73–46, Ottawa

Arigo, R., S.E. Howe, and T. Webb III. 1982. Computer programs for climatic calibration of pollen data. In Berglund 1982: 79–109

Barry, R.G. 1982. Variability of the climate system. American Quaternary Association, *Abstracts Volume* 2–3, Seattle

Barry, R.G., and F.K. Hare. 1974. Arctic climate. In J.D. Ives and R.G. Barry, eds, *Arctic and Alpine Environments* (Methuen, London): 17–54

Berger, A. 1978. Long-term variations of caloric insolation resulting from the earth's orbital elements. *Quaternary Research 9:* 139–67

– 1981. Astronomical theory of paleoclimates. In A. Berger, ed., *Climatic Variations and Variability: Facts and Theories* (Reidel, London): 501–26

Berggren, W.A. 1972. A Cenozoic time scale: some implications for regional geology and paleobiogeography. *Lethaia,* 5: 195–215

Berglund, B.E., ed. 1979a. *Palaeohydrological Changes in the Temperate Zone in the Last 15,000 Years.* Project Guide, vols 1 and 2, Department of Quaternary Geology, Lund University, Lund

Berglund, B.E. 1979b. Pollen Analysis. In Berglund 1979a, vol. 2: 133–67

– ed. 1982. *Palaeohydrological Changes in the Temperate Zone in the Last 15,000 Years.* Project Guide, vol. 3, Department of Quaternary Geology, Lund University, Lund

Beschel, R.E. 1970. The diversity of tundra vegetation. In W.A. Fuller and P.G. Kevan, eds, *Productivity and Conservation in Northern Circumpolar Lands* (International Union for Conservation of Nature and Natural Resources Publications, Morges, new series 16): 85–92

Billings, W.D. 1973. Arctic and alpine vegetations: similarities, differences, and suscepti-
bility to disturbance. *Bioscience*, 23: 697–704

– 1974. Plant adaptations to cold summer climates. In J.D. Ives and R.G. Barry, eds,
Arctic and Alpine Environments (Methuen, London): 403–43

Bird, C.D., G.W. Scotter, and W.G. Steere. 1977. Bryophytes from the area drained by
the Peel and Mackenzie Rivers, Yukon and Northwest Territories, Canada. *Canadian
Journal of Botany*, 55: 2879–2918

Birks, H.H. 1980. Plant macrofossils in Quaternary lake sediments. *Archiv für Hydrobi-
ologie, Ergelbnisse der Limnologie*, 15: 1–60

Birks, H.J.B. 1973. *Past and Present Vegetation on the Isle of Skye: A Palaeoecological
Study.* Cambridge University Press, Cambridge

– 1979. Guidelines for numerical treatment of data. In Berglund 1979a, vol. 1: 99–123

– 1981. The use of pollen analysis in the reconstruction of past climates: a review. In
T.M.L. Wigley, M.J. Ingram, and G. Farmer, eds, *Climate and History* (Cambridge
University Press, Cambridge): 111–37

Birks, H.J.B., and H.H. Birks. 1980. *Quaternary Palaeoecology.* Arnold, London

Birks, H.J.B., and B. Huntley. 1983. *An Atlas of Past and Present Pollen Maps for
Europe: 0–13000 Years Ago.* Cambridge University Press, Cambridge

Birks, H.J.B., and S.M. Peglar. 1980. Identification of *Picea* pollen of Late-Quaternary
age in eastern North America: a numerical approach. *Canadian Journal of Botany*, 58:
2043–58

Biske, S.F. 1974. *Correlation of Tertiary Non-marine Deposits in Alaska and Northeastern
Asia.* American Association of Petroleum Geologists, Memoir 19: 239–45

Black, R.A. 1979. The reproductive biology of *Picea mariana* (Mill.) BSP. at treeline.
PhD thesis, University of Alberta, Edmonton

Black, R.A., and L.C. Bliss. 1978. Recovery sequence of *Picea mariana / Vaccinium uliginosum* forests after burning near Inuvik, Northwest Territories, Canada. *Canadian
Journal of Botany*, 56: 2020–30

– 1980. Reproductive ecology of *Picea mariana* (Mill.) BSP at treeline near Inuvik,
Northwest Territories, Canada. *Ecological Monographs*, 50: 331–54

Bliss, L.C. 1975. Tundra grasslands, herblands and shrublands and the role of herbivores. *Geoscience and Man*, 10: 51–79

Bliss, L.C., and J.E. Cantlon. 1957. Succession in river alluvium in northern Alaska.
American Midland Naturalist, 58: 452–69

Bliss, L.C., and J. Richards. 1982. Present day arctic vegetation as a predictive tool for
the arctic steppe – mammoth biome. In Hopkins et al. 1982: 241–57

Bostock, H.S. 1948. *Physiography of the Canadian Cordillera with Special Reference to
the Area North of the Fifty-fifth Parallel.* Geological Survey of Canada, Memoir 247,
Ottawa

Bradshaw, R.H.W. 1981. Quantitative reconstruction of local woodland vegetation
using pollen analysis from a small basin in Norfolk, England. *Journal of Ecology*, 69:
949–55

Briggs, N.D., and J.A. Westgate. 1978. Fission-track age of tephra marker beds in
Beringia. American Quaternary Association, 5th Biennial Meeting, Edmonton,
Alberta, *Abstracts*: 190

Britton, M.E. 1967. Vegetation of the arctic tundra. In H.P. Hansen, ed., *Arctic Biology*,
2nd ed. (Oregon State University Press, Corvallis): 67–130

Brown, R.J.E. 1967. Permafrost in Canada. Map 1246A, Geological Survey of Canada, Ottawa

Brubaker, L.B., H.L. Garfinkel, and M.E. Edwards. 1983. A Late-Wisconsin and Holocene vegetation history from the central Brooks Range: implications for Alaskan paleoecology. *Quaternary Research*, 19: in press

Bryson, R.A. 1966. Air masses, streamlines and the Boreal Forest. *Geographical Bulletin*, 8: 228–69

Budyko, M.I. 1982. *The Earth's Climate: Past and Future.* Academic Press, New York

Burns, B.M. 1973. *The Climate of the Mackenzie Valley–Beaufort Sea*, vol. 1. Climatological Studies No. 24, Environment Canada, Toronto

– 1974. *The Climate of the Mackenzie Valley–Beaufort Sea*, vol. 2. Climatological Studies no. 24, Environment Canada, Toronto

Canby, T.Y. 1979. The search for the first Americans. *National Geographic*, 156: 330–63

Carter, R.N., and S.D. Prince. 1981. Epidemic models used to explain biogeographical distribution limits. *Nature*, 293: 644–5

Churchill, E.D. 1955. Phytosociological and environmental characteristics of some plant communities in the Umiat region of Alaska. *Ecology*, 36: 606–27

Cinq-Mars, J. 1979. Bluefish Cave I: a Late Pleistocene eastern Beringian cave deposit in the northern Yukon. *Canadian Journal of Archaeology*, 3: 1–32

Clebsch, E.E.C., and R.E. Shanks. 1968. Summer climatic gradient and vegetation near Barrow, Alaska. *Arctic*, 21: 161–71

CLIMAP 1976. The surface of the Ice-Age earth. *Science*, 191: 1131–7

Colinvaux, P.A. 1964. The environment of the Bering Land Bridge. *Ecological Monographs*, 34: 297–329

– 1967. Quaternary vegetational history of arctic Alaska. In D.M. Hopkins, ed., *The Bering Land Bridge* (Stanford University Press, Stanford): 207–31

– 1980. Arctic steppe-tundra. *Quarterly Review of Archaeology*, 1: 2

Corns, I.G.W. 1974. Arctic plant communities east of the Mackenzie Delta. *Canadian Journal of Botany*, 52: 1731–45

Cwynar, L.C. 1980. A late-Quaternary vegetation history from Hanging Lake, northern Yukon. PhD thesis, University of Toronto

– 1982. A late-Quaternary vegetation history from Hanging Lake, Northern Yukon. *Ecological Monographs*, 52: 1–24

Cwynar, L.C., E. Burden, and J.H. McAndrews. 1979. An inexpensive sieving method for concentrating pollen and spores from fine-grained sediments. *Canadian Journal of Earth Sciences*, 16: 1115–20

Cwynar, L.C., and J.C. Ritchie. 1980. Arctic steppe-tundra: A Yukon perspective. *Science*, 208: 1375–7

Davis, M.B. 1963. On the theory of pollen analysis. *American Journal of Science*, 261: 897–912

– 1967. A method for determination of absolute pollen frequency. In B. Hummel and D.M. Raup, eds, *Handbook of Paleontological Techniques* (W.H. Freeman, San Francisco)

– 1976. Pleistocene biogeography of temperate deciduous forests. *Geoscience and Man*, 13: 13–26

– 1981. Quaternary history and the stability of forest communities. D.C. West, H.H.

Shugart, and D.B. Botkin, eds, *Forest Succession: Concepts and Application* (Springer-Verlag, New York): 132–53

Deevey, E.S. 1969. Coaxing history to conduct experiments. *Bioscience*, 19: 40–3

Delcourt, H.R., P.A. Delcourt, and T. Webb III. 1983. Dynamic plant ecology: the spectrum of vegetational change in space and time. *Quaternary Science Reviews*, in press

Delorme, L.D., S.C. Zoltai, and L.L. Kalas. 1977. Freshwater shelled invertebrate indicators of paleoclimate in north-western Canada during late glacial times. *Canadian Journal of Earth Sciences*, 14: 2029–46

De Lumley, H., J.-C. Miskovsky, J. Renault-Miskovsky, and Jean-Pierre Gerber. 1973. Le Würmien Ancien-Etudes Françaises sur le Quaternaire. *9th International Quaternary Association Congress Proceedings*, Christchurch, New Zealand

Dingman, S.L., R.G. Barry, G. Weller, C. Benson, E.F. LeDrew, and C.W. Goodwin. 1980. Climate, snowcover, microclimate and hydrology. In J. Brown, P.C. Miller, L.L. Tiezen, and F.L. Bunnell, eds, *An Arctic Ecosystem* (Dowden, Hutchinson and Ross, Stroudsburg, Pa.): 30–71

Dobbs, R.C. 1976. White spruce seed dispersal in central British Columbia. *Forestry Chronicle*, 52: 225–8.

Drew, J.V., and R.E. Shanks. 1965. Landscape relationships of soils and vegetation in the forest-tundra ecotone, Upper Firth River Valley, Alaska-Canada. *Ecological Monographs*, 35: 285–306

Drury, W.H. 1969. Plant persistence in the Gulf of St. Lawrence. In K.N.H. Greenidge, ed., *Essays in Plant Geography and Ecology* (Nova Scotia Museum, Halifax): 105–48

Elliott, D.L. 1979. The stability of the Northern Canadian Tree Limit: current regenerative capacity. PhD thesis, University of Colorado, Boulder

Faegri, K., and J. Iversen. 1975. *Textbook of Pollen Analysis*, 3rd ed. Munksgaard, Copenhagen

Foote, J. 1979. General patterns of forest succession in interior Alaska. *Proceedings of XIIIth Alaska Science Conference*, Fairbanks: 20–1

Fowells, H.A. 1965. *Silvics of Forest Trees of the United States.* U.S. Department of Agriculture, Handbook 271

Fredskild, B. 1969. A postglacial standard pollen diagram from Peary Land, North Greenland. *Pollen et Spores*, 11: 573–83

Freitag, H. 1977. The periglacial, late-glacial and early postglacial vegetations of Zeribar and their present-day counterparts. *Palaeohistoria*, 19: 87–95

French, H.M. 1976. *The Periglacial Environment.* Longman, New York

Fyles, J.G., J.A. Heginbottom, and V.N. Rampton. 1972. Quaternary geology and geomorphology, Mackenzie Delta to Hudson Bay. *24th International Geological Congress, Montreal, Excursion A30*: 23

Gagnon, R. 1982. Fluctuations holocènes de la limite des forets, Rivière aux Feuilles, Québec nordique: une analyse macrofossile. Thèse de doctorat, Université Laval

Garfinkel, H.L., and L.B. Brubaker. 1980. Modern climate-tree-growth relationships and climatic reconstruction in sub-arctic Alaska. *Nature*, 286: 872–4

Gilbert, H. and S. Payette. 1982. Ecologie des populations d'Aulne vert (*Alnus crispa* (Ait.) Pursh) à la limite des forets, Québec Nordique. *Géographie physique et quaternaire*, 36: 109–24

Gill, D. 1971. Vegetation and environment in the Mackenzie River Delta; a study in subarctic ecology. Unpublished PhD thesis, University of British Columbia, Vancouver

– 1972. The point bar environment in the Mackenzie River Delta. *Canadian Journal of Earth Sciences*, 9: 1382–3

– 1973a. Ecological modifications caused by the removal of tree and shrub canopies in the Mackenzie Delta. *Arctic*, 26: 95–111

– 1973b. A spatial correlation between plant distribution and unfrozen ground within a region of discontinuous permafrost. In *Permafrost, Second International Conference* (National Academy of Sciences, Washington): 105–13

– 1973c. Modification of northern alluvial habitats by river development. *The Canadian Geographer* 17: 138–53

– 1975. Influence of white spruce trees on permafrost-table microtopography, Mackenzie River Delta. *Canadian Journal of Earth Sciences*, 12: 263–72

Giterman, R.E. 1980. The history of Pliocene and Pleistocene vegetation in the western part of Beringia. *IV International Palynology Conference Proceedings*, Lucknow, 3: 134–9

Giterman, R.E., A.V. Sher, and J.V. Matthews. 1982. Comparison of the development of tundra-steppe environments in West and East Beringia: Pollen and macrofossil evidence from key sections. In Hopkins et al. 1982: 43–73

Gjaerevoll, O. 1954. Kobresieto-Dryadion in Alaska. *Nytt Magazin Botanik*, 3: 51–5

– 1958. Botanical investigations in Central Alaska, especially in the White Mountains. Part I. Pteridophytes and monocotyledones. *Det Kungliga Norske Videnskabers Selskabs Skrifter*, 5: 1–74

– 1963. Botanical investigations in Central Alaska, especially in White Mountains. Part II. Dicotyledones Salicaceae-Umbelliferae. *Det Kungliga Norske Videnskabers Selskabs Skrifter*, 4: 1–115

– 1967. Botanical investigations in Central Alaska, especially in White Mountains. Part III. Sympetalae. *Det Kungliga Norske Videnskabers Selskabs Skrifter* 10: 1–63

– 1980. A comparison between the alpine plant communities of Alaska and Scandinavia. *Acta Phytogeographica Suecica*, 68: 83–8

Gleason, H.A. 1926. The individualistic concept of the plant association. *Bulletin of the Torrey Botanical Club*, 44: 463–81

Gleason, H.A., and A. Cronquist. 1964. *The Natural Geography of Plants*. Columbia, New York

Godwin, H. 1975. *The History of the British Flora*, 2nd ed. Cambridge University Press, Cambridge

Gorodkov, B.N. 1958. The vegetation and soils of Wrangell Island. *Vegetation of Far North of the USSR*, 3: 5–58

Gould, S.J. 1975. Is uniformitarianism necessary? *American Journal of Science*, 5: 223–8

Green, D.G. 1982. Fire and stability in the post-glacial forests of southwest Nova Scotia. *Journal of Biogeography*, 9: 29–40

Green, G.C. 1981. Time series and post-glacial forest ecology. *Quaternary Research*, 15: 265–77

Griggs, R.F. 1934. The problem of Arctic vegetation. *Journal of the Washington Academy of Science*, 24: 153–75

Guthrie, R.D. 1968. Paleoecology of the large mammal community in interior Alaska during the Late Pleistocene. *American Midland Naturalist*, 79: 346–63

– 1982. Mammals of the Mammoth-Steppe as paleoenvironmental indicators. In Hopkins et al. 1982: 307–26

Hamilton, T.D. 1979. *Radiocarbon Dates and Quaternary Stratigraphic Sections, Philip Smith Mountains Quadrangle, Alaska.* U.S. Geological Survey, Open-File Report 79–866

– 1982. A Late Pleistocene glacial chronology for the southern Brooks Range: Stratigraphic record and regional significance. *Geological Society of America Bulletin*, 93: 700–16

Hanson, H.C. 1950. Vegetation and soil profiles in some solifluction and mound areas in Alaska. *Ecology*, 31: 606–30

– 1951. Characteristics of some grassland, marsh, and other plant communities in western Alaska. *Ecological Monographs*, 21: 317–78

– 1953. Vegetation types in northwestern Alaska and comparisons with communities in other arctic regions. *Ecology*, 34: 111–40

Hare, F.K. 1968. The Arctic. *Quarterly Journal of the Royal Meteorological Society*: 94

Hare, F.K., and J.E. Hay. 1974. The climate of Canada and Alaska. In H.E. Landsberg, ed., *Climates of North America*, vol. II, *World Survey of Climatology* (Elsevier, New York): 49

Hare, F.K., and J.C. Ritchie. 1972. The boreal bioclimates. *Geographical Review*: 334–65

Harington, C.R. 1978. *Quaternary Vertebrate Faunas of Canada and Alaska and Their Suggested Chronological Sequence.* Syllogeus, 15

– 1980. Pleistocene Saiga Antelopes in North America and their Paleoenvironmental Implications. In W.C. Mahaney, ed., *Quaternary Paleoclimate* (Geoabstracts, Norwich): 193–225

Harper, J.C. 1982. After description. In *The Plant Community as a Working Mechanism* (British Ecological Society, Special Publications 1, Blackwell, Oxford): 11–25

Hettinger, L.A., A. Janz, and R.W. Wein. 1973. *Vegetation of the Northern Yukon.* Arctic Gas, Biological Report Series 1, Calgary

Hills, L.V. 1975. Late Tertiary floras of Arctic Canada; an interpretation. *Proceedings Circumpolar Conference on Northern Ecology* (Ottawa)

Hills, L.V., and J.V. Matthews Jr. 1974. A preliminary list of fossil plants from the Beaufort Formation, Meighen Island, District of Franklin. *Geological Survey of Canada, Paper 74-1B*: 224–6

Hills, L.V., and R.T. Ogilvie. 1970. *Picea banksii* n.sp. Beaufort Formation (Tertiary) northwestern Banks Island, arctic Canada. *Canadian Journal of Botany*, 48: 457–64

Hoffmann, R.S. 1974. Terrestrial vertebrates. In J.D. Ives and R.G. Barry, eds, *Arctic and Alpine Environments* (Methuen, London): 475–568

Hopkins, D.M. 1972. The paleogeography and climatic history of Beringia during late Cenozoic time. *Internord*, 12: 121–50

– 1979. Landscape and climate of Beringia during Late Pleistocene and Holocene time. In W.S. Laughlin and A.B. Harper, eds, *The First Americans: Origins, Affinities and Adaptations* (Fischer-Verlag, New York): 15–41

Hopkins, D.M., J.V. Matthews, C.E. Schweger, and S.B. Young, eds. 1982. *Paleoecology of Beringia.* Academic Press, New York

Hopkins, D.M., J.V. Matthews, J.A. Wolfe, and M.L. Silberman. 1971. A Pliocene flora and insect fauna from the Bering Strait region. *Paleoclimatology, Paleogeography and Paleoecology*, 9: 211–31

Hopkins, D.M., and R.S. Sigafoos. 1951. Frost action and vegetation patterns on Seward Peninsula, Alaska. *U.S. Geological Survey, Bulletin 974–C*: 51–100

Hopkins, D.M., P.A. Smith, and J.V. Matthews Jr. 1981. Dated wood from Alaska and the Yukon: implications for forest refugia in Beringia. *Quaternary Research*, 15: 217–49

Howe, S., and T. Webb III. 1983. Calibrating pollen data in climatic terms: improving the methods. *Quaternary Science Reviews*, in press

Hubert, B.A. 1977. Estimated productivity of muskox on Truelove Lowland. In L.C. Bliss, ed., *Truelove Lowland, Devon Island, Canada, a High Arctic Ecosystem* (University of Alberta Press, Edmonton): 467–91

Hughes, O.L. 1969. Pleistocene stratigraphy, Porcupine and Old Crow Rivers, Yukon Territory. *Geological Survey of Canada, Paper 69–1*: 209–12

– 1972. *Surficial Geology of Northern Yukon Territory and Northwestern District of Mackenzie, Northwest Territories*. Geological Survey of Canada, Report 69–36 and map 1319A

Hughes, O.L., C.R. Harington, J.A. Janssens, J.V. Matthews, R.E. Morlan, N.W. Rutter, and C.E. Schweger. 1981. Upper Pleistocene stratigraphy, paleoecology and archaeology of the Northern Yukon Interior, Eastern Beringia. I. Bonnet Plume Basin. *Arctic*, 34: 329–65

Hughes, O.L., and J. Terasmae. 1963. SIPRE ice-corer for obtaining samples from permanently frozen bogs. *Arctic*, 16: 270–2

Hultén, E. 1927–8. *Flora of Kamtchatka and the Adjacent Islands*. Kungliga Svenska Vetenskapsakadamiens, Stockholm

– 1937. *Outline of the History of Arctic and Boreal Biota during the Quaternary Period*. Bokforlagsaktiebologet Thule, Stockholm

– 1962. Flora and vegetation of Scammon Bay, Bering Sea Coast, Alaska. *Svenska Botaniska Tidsskift*, 56: 36–54

– 1968. *Flora of Alaska and Neighboring Territories*. Stanford University Press, Stanford

Hyvärinen, H., and J.C. Ritchie. 1975. Pollen stratigraphy of Mackenzie pingo sediments, N.W.T, Canada. *Arctic and Alpine Research*, 7: 261–72

Inglis, J.T. 1975. Vegetation and reindeer-range relationships in the forest tundra transition zone, Sitidgi Lake area, N.W.T. Unpublished MSc thesis, Carleton University, Ottawa

Irving, W.N., and C.R. Harington. 1973. Upper Pleistocene radiocarbon-dated artefacts from the northern Yukon. *Science*, 179: 335–40

Iversen, J. 1964. Plant indicators of climate, soil and other factors during the Quaternary. *Report, VIth International Congress on Quaternary Science*, Section 2: 421–8

Ives, J.W. 1977. Pollen separation of three North American birches. *Arctic and Alpine Research*, 9: 73–80

Jacobs, J.D., and C.Y.Y. Leung. 1980. Paleoclimatic implications of topoclimatic diversity in arctic Canada. In W.C. Mahaney, ed., *Quaternary Paleoclimate* (Geoabstracts, Norwich): 63–75

Jacobson, G.L., and R.H.W. Bradshaw. 1981. The selection of sites for paleovegetational studies. *Quaternary Research*, 16: 80–96

Jakimchuk, R.D., and D.R. Carruthers. 1980. *Caribou and Muskoxen on Victoria Island, N.W.T.* Polar Gas Project Report, Toronto

Jefferies, R.L. 1977. The vegetation of salt marshes at some coastal sites in arctic North America. *Journal of Ecology*, 65: 661-72

Johansen, F. 1924. *Report on the Canadian Arctic Expedition 1913-1918*, vol. 5, part C: 1c-86c (Ottawa)

Johnson, A.W., and J.G. Packer. 1967. Distribution, ecology and cytology of the Ogotoruk Creek flora and the history of Beringia. In D.M. Hopkins, ed., *The Bering Land Bridge* (Stanford University Press, Stanford): 245-65

Johnson, A.W., L.A. Viereck, R.E. Johnson, and H. Melchior. 1966. Vegetation and flora of the Cape Thompson-Ogotoruk Creek area, Alaska. In N.J. Wilimovsky and J.N. Wolfe, eds, *Environment of the Cape Thompson Region, Alaska* (U.S. Atomic Energy Commission, Division of Technical Information, PNE-481): 277-354

Jopling, A.V., W.N. Irving, and B.F. Beebe. 1981. Stratigraphic, sedimentological and faunal evidence for the occurrence of Pre-Sangamonian artefacts in northern Yukon. *Arctic*, 34: 3-33

Katz, N.Y., S.V. Katz, A.V. Sher, and R.E. Giterman. 1970a. On the Lower Pleistocene flora of the eastern part of the coastal plain (in Russian). In A.E. Tolmatchev, ed. *Cenozoic History of the Polar Basin* (Academy of Science, USSR): 74-5

– 1970b. The lower Pleistocene flora of the eastern part of the coastal plain. In A.E. Sher, ed., *Severniy Ledovitiy Okean i ego poberezh'e v Kaynozoe* [The Northern Arctic Ocean and Its History in the Cenozoic] (Leningrad): 483-3

Kay, P.A. 1979. Multivariate statistical estimates of Holocene vegetation and climate change, forest-tundra transition zone, Northwest Territories, Canada. *Quaternary Research*, 11: 125-40

Kerfoot, D.E. 1969. The geomorphology and permafrost conditions of Garry Island, N.W.T. Unpublished PhD thesis, University of British Columbia, Vancouver

Krattinger, K. 1975. Genetic mobility in *Typha*. *Aquatic Botany*, 1: 57-70

Kubiack, H. 1982. Morphological characters of the mammoth: an adaptation to the arctic-steppe environment. In Hopkins et al. 1982: 281-9

Kuc, M. 1974. Fossil flora of the Beaufort Formation, Meighen Island, Northwest Territories. *Geological Survey of Canada, Report of Activities*, Paper 74-1: 193-5

Kukla, G.J. 1981. Pleistocene climates on land. In A. Berger, ed., *Climatic Variations and Variability: Facts and Theories* (Reidel, London): 207-32

Lamb, H.H. 1977. *Climate, Present, Past and Future*, vol. 2. Methuen, London

Lamb, H.H., and A. Woodroffe. 1970. Atmospheric circulation during the last Ice Age. *Quaternary Research*, 1: 29-59

Lambert, J.D.H. 1968. The ecology and successional trends of tundra plant communities in the Low Arctic Subalpine zone of the Richardson and British Mountains of the Canadian western arctic. PhD thesis, University of British Columbia, Vancouver

– 1972. Plant succession on tundra mudflows: preliminary observations. *Arctic*, 25: 99-106

Larsen, J.A. 1980. *The Boreal Ecosystem*. Academic Press, New York

Lavrenko, E.M., and V.B. Sochava. 1956. *The Vegetation Cover of the USSR* Scientific Edition, Academy of Sciences, vol. 1, Moscow

Lichti-Federovich, S. 1970. The pollen stratigraphy of a dated section of late-Pleistocene lake sediment from central Alberta. *Canadian Journal of Earth Sciences*, 7: 938-45

- 1972. Pollen stratigraphy of a sediment core from Alpen Siding, Alberta. *Geological Survey of Canada, Report of Activities*, Paper 72-1B: 113-15
- 1973. Palynology of six sections of late Quaternary sediments from the Old Crow River, Yukon Territory. *Canadian Journal of Botany*, 51: 533-64
- 1974. *Palynology of Two Sections of Late Quaternary Sediments from the Porcupine River, Yukon Territory.* Geological Survey of Canada, Paper 74-23

Lichti-Federovich, S., and J.C. Ritchie, 1968. Recent pollen assemblages from the Western Interior of Canada. *Review of Palaeobotany and Palynology*, 7: 297-344

Livingstone, D.A. 1955a. Some pollen profiles from arctic Alaska. *Ecology*, 36: 587-600
- 1955b. A lightweight piston sampler for lake deposits. *Ecology*, 36: 137-9

Löve, A., and D. Löve. 1974. Origin and evolution of the arctic and alpine floras. In J.D. Ives and R.G. Barry, eds, *Arctic and Alpine Environments* (Methuen, London): 571-603

MacDonald, G.M. 1982. Late Quaternary paleoenvironments of the Morley Flats and Kananaskis Valley of southewestern Alberta. *Canadian Journal of Earth Sciences*, 19: 23-35

Mackay, G.A., B.F. Findlay, and H.A. Thompson. 1975. A climatic perspective of tundra areas. In *Proceedings of Circumpolar Conference on Northern Ecology* (National Research Council of Canada, Ottawa): 10-33

Mackay, J.R. 1963. *The Mackenzie Delta Area, N.W.T.* Ottawa, Geographical Branch, Memoir 8
- 1967. Permafrost depths, Lower Mackenzie Valley, Northwest Territories. *Arctic*, 20: 21-6
- 1971. The origin of massive icy beds in permafrost, Western Arctic Coast, Canada. *Canadian Journal of Earth Sciences*, 8: 397-422
- 1972. The world of underground ice. *Annals of the Association of American Geographers*, 62: 1-22
- 1973. The growth of pingos, Western Arctic Coast, Canada. *Canadian Journal of Earth Sciences*, 10: 979-1004
- 1975a. Relict ice wedges, Pelly Island, N.W.T. *Geological Survey of Canada, Paper 75-1A*: 469-70
- 1975b. The stability of permafrost and recent climatic change in the Mackenzie Valley, N.W.T. *Geological Survey of Canada, Paper 75-1B*: 173-6
- 1976. The age of Ibyuk Pingo, Tuktoyaktuk Peninsula, District of Mackenzie. *Geological Survey of Canada, Paper 76-1B*: 59-60
- 1978. Freshwater shelled invertebrate indicators of paleoclimate in northwestern Canada during late glacial times: discussion. *Canadian Journal of Earth Sciences*, 15: 461-2
- 1979. Pingos of the Tuktoyaktuk Peninsula area, Northwest Territories. *Géographie physique et quaternaire*, 33 (1): 3-61

Mackay, J.R., and D.K. Mackay. 1974. Snow cover and ground temperatures, Garry Island, N.W.T. *Arctic*, 27: 287-96

Mackay, J.R., V.N. Rampton, and J.G. Fyles. 1972. Relict Pleistocene permafrost, Western Arctic, Canada. *Science*, 176: 1321-3

Mackay, J.R., and J. Terasmae. 1963. Pollen diagrams in the Mackenzie Delta Area, N.W.T. *Arctic*, 16: 229-38

Macpherson, A.H. 1961. On the abundance and distribution of certain mammals in the Western Canadian Islands in 1958-59. *The Arctic Circular*, 14: 1

Maher, L.J. Jr. 1972. Absolute pollen diagram of Redrock Lake, Boulder County, Colorado. *Quaternary Research*, 2: 531–53

– 1981. Statistics for microfossil concentration measurements employing samples spiked with marker grains. *Review of Palaeobotany and Palynology*, 32: 153–91

Manning, T.H. 1960. *The Relationship of the Peary and Barren-Ground Caribou.* Arctic Institute of North America Technical Paper no. 4: 52

Martin, P.S. 1967. Prehistoric overkill. In P.S. Martin and H.E. Wright, eds., *Pleistocene Extinctions* (Yale University Press, New Haven): 76–120

– 1982. The pattern and meaning of Holarctic mammoth extinction. In Hopkins et al. 1982: 399–408

Matthews, J.V. Jr. 1974. Wisconsin environment of interior Alaska: pollen and macrofossil analysis of a 27 m core from Isabella Basin (Fairbanks, Alaska). *Canadian Journal of Earth Sciences*, 11: 812–41

– 1975a. Insects and plant macrofossils from two Quaternary exposures in the Old Crow–Porcupine region, Yukon Territory, Canada. *Arctic and Alpine Research*, 7: 249–59

– 1975b. Incongruence of macrofossil and pollen evidence, a case from the Late Pleistocene of the Northern Yukon coast. *Geological Survey of Canada, Paper 75-1*: 139–46

– 1976. Arctic steppe: an extinct biome. *Abstracts of the Fourth Biennial Conference of the American Quaternary Association*, Tempe: 73–7

– 1980. *Paleoecology of John Klondike Bog, Fisherman Lake Region, Southwest District of Mackenzie.* Geological Survey of Canada, Paper 80–22

– 1982. East Beringia during Late Wisconsin time: A review of the biotic evidence. In Hopkins et al. 1982: 127–50

Maxwell, J.B. 1980. *The Climate of the Canadian Arctic Islands and Adjacent Waters*, vol. 1. Environment Canada, Ottawa

McCulloch, D., and D.M. Hopkins. 1966. Evidence for an early recent warm interval in northwestern Alaska. *Geological Society of America, Bulletin* 77: 1089–1108

McIntosh, R.P. 1976. Ecology since 1900. In B.J. Taylor and T.J. White, eds, *Issues and Ideas in America.* (University of Oklahoma Press, Norman): 253–73

Meltzer, D.J., and J.I. Mead. 1983. The timing of Late Pleistocene mammalian extinctions in North America. *Quaternary Research*, 19: 130–5

Miller, F.L., R.M. Russell, and A. Gunn. 1977. *Peary Caribou and Muskoxen on Western Queen Elizabeth Islands, N.W.T. 1972–74.* Canadian Wildlife Service, Series no. 40

Mitchell, G.F. 1971. Fossil pingos in the south of Ireland. *Nature*, 230 (March 5): 43–4

Monteith, J.L. 1966. Photosynthesis and transpiration of crops. *Experimental Agricultural Review*, 2: 1–14

– 1981. Climatic variation and the growth of crops. *Quarterly Journal of the Royal Meteorological Society*, 107: 749–74

Morgan, A.V. 1972. Late Wisconsin ice-wedge polygons near Kitchener, Ontario, Canada. *Canadian Journal of Earth Sciences*, 9, 607–17

Morlan, R.E. 1980. *Taphonomy and Archaeology in the Upper Pleistocene of the Northern Yukon Territory: A Glimpse of the Peopling of the New World.* Archaeological Survey of Canada, Paper 94, Ottawa

Mosquin, T. 1966. Reproductive specialization as a factor in the evolution of the Canadian flora. In R.L. Taylor and R.A. Ludwig, eds., *The Evolution of Canada's Flora* (University of Toronto Press, Toronto): 43–65

Müller-Beck, H. 1982. Upper Pleistocene man in northern Eurasia and the Steppe mammoth biome. In Hopkins et al. 1982: 327–52

Muratova, M.V. 1973. *Historical Development of Vegetation and Climate in Southeast Chukotka during the Neogene-Pleistocene.* 'Nauka,' Moscow

Murray, D.F. 1978. Vegetation, floristics and phytogeography of northern Alaska. In L.L. Tieszen, ed., *Vegetation and Production Ecology of an Alaskan Arctic Tundra* (Ecological Studies, 29, Springer-Verlag, New York): 19–36

– 1980. Balsam poplar in arctic Alaska. *Canadian Journal of Anthropology*, 1: 29–32

– 1981. The role of Arctic refugia in the evolution of the Arctic vascular flora – a Beringian perspective. In G.G.E. Scudder and J.L. Reveal, eds, *Evolution Today* (University of British Columbia Press, Vancouver): 11–20

Murray, D.F., and A.R. Batten. 1977. A provisional classification of Alaskan tundra. Unpublished report, University of Alaska and Pacific Northwest Forest and Range Experimental Station

National Atlas of Canada, The (4th edition, revised). Macmillan, Ottawa 1974

Nichols, H. 1976. Historical aspects of the northern Canadian treeline. *Arctic*, 29: 38–47

Nienstaedt, M., and A. Teich. 1971. *The Genetics of White Spruce.* U.S. Department of Agriculture, Forest Service, Research Paper W0-15

Norris, D.K. 1973. *Tectonic Styles of Northern Yukon Territory and Northwestern District of Mackenzie, Canada.* American Association of Petroleum Geologists, Memoir 19: 23–40

Norris, D.K., and C.J. Yorath. 1981. The North American Plate from the Arctic Archipelago to the Romanzof Mountains. In A.E.M. Nairn, M. Churkin, and F.G. Stehli, eds, *The Arctic Ocean*, vol. 5 of the Ocean Basins and Margins (Plenum, New York): 37–103

Norris, G. 1982. Spore-pollen evidence for early Oligocene high-latitude cool climatic episode in northern Canada. *Nature*, 297: 387–9

Ohmura, A. 1982. Climate and energy balance in the arctic tundra. *Journal of Climatology*, 2: 65–84

Olsson, I.V. 1974. Some problems in connection with the evaluation of C^{14} dates. *Geologiska föreningens Förhandlingan*, Stockholm, 96: 311–20

Oswald, E.T., and J.P. Senyk. 1977. *Ecoregions of Yukon Territory.* Canadian Forestry Service, Victoria

Ovenden, L.E. 1981. Vegetation history of a polygonal peatland, Old Crow Flats, Northern Yukon. MSc Thesis, Department of Botany, University of Toronto

– 1982. Vegetation history of a polygonal peatland, northern Yukon. *Boreas*, 11: 209–24

Owens, J.N., and M. Molder. 1977. Bud development in *Picea glauca.* II. Cone differentiation and early development. *Canadian Journal of Botany*, 55: 2746–60

Owens, J.N., M. Molder, and H. Langer, 1977. Bud development in *Picea glauca.* I. Annual growth cycle of vegetative buds and shoot elongation as they relate to date and temperature sums. *Canadian Journal of Botany*, 55: 2728–45

Parker, G.R. 1978. *The Diets of Muskoxen and Peary Caribou on Some Islands in the Canadian High Arctic.* Occasional Paper 35, Canadian Wildlife Service, Ottawa

Parker, G.R., and R.K. Ross. 1976. Summer habitat use by muskoxen (*Ovibos moschatus*) and Peary caribou (*Rangifer tarandus pearyi*) in the Canadian High Arctic. *Polarforschung*, 46: 12–25

Parson, R.W., and I.C. Prentice. 1981. Statistical approaches to *R*-values and the pollen-vegetation relationship. *Review of Palaeobotany and Palynology*, 32: 127–52

Patterson, W.A., B.F. Wilson, and J.F. O'Keefe. 1983. The ecology of Alder (*Alnus viridis* Furlow) at its range limit in Northwest Alaska. *Abstracts of the 12th Annual Arctic Workshop*, University of Massachusetts, Amherst: 50–1

Payette, S., and R. Gagnon. 1979. Tree-line dynamics in Ungava Peninsula. *Holarctic Ecology*, 2: 239–48

Pearce, G.W., J.A. Westgate, and S. Robertson. 1982. Magnetic reversal history of Pleistocene sediments at Old Crow, northwestern Yukon Territory. *Canadian Journal of Earth Sciences*, 19: 919–29

Penman, H.L. 1948. Natural evaporation from open water, base soil and grass. *Proceedings of the Royal Society*, A193: 120–45

Peterson, E.B., E.D. Kabzems, and V.M. Levson. 1980. *Terrain and Vegetation along the Victoria Island Portion of a Polar Gas Combine Pipeline System.* Polar Gas Report, Toronto

Péwé, T.L. 1965. *Central and South Central Alaska: Guidebook for Field Conference F,* International Quaternary Association 7th Congress, Boulder

Pissart, A. 1963. Les traces de 'pingos' du Pays de Galles (Grande Bretagne) et du Plateau des Hautes Fagnes (Belgique). *Annales géomorphologiques*, 7: 147–65

Porsild, A.E. 1938. Earth mounds in unglaciated arctic northwestern America. *Geographical Review*, 28: 46–58

– 1955. *Botany of Southeastern Yukon Adjacent to the Canol Road.* National Museum of Canada, Bulletin 121

Porsild, A.E., and W.J. Cody. 1980. *Vascular Plants of Continental Northwest Territories, Canada.* National Museum of Natural Sciences, Ottawa

Prentice, I.C. 1982. Calibration of pollen spectra in terms of species abundance. In Berglund 1982: 25–51

– 1983. Postglacial climatic change: vegetation dynamics and the pollen record. *Progress in Physical Geography*, 7: in press

Prest, V.K. 1970. Quaternary geology of Canada. In R.J.E. Douglas, ed., *Geology and Economic Minerals of Canada* (Department of Energy, Mines and Resources, Ottawa): 676–764

Racine, C.H. 1974. Vegetation of the Chukchi-Imuruk area. MS, Chukchi-Imuruk Biological Survey, Cooperative Parks Studies Unit, University of Alaska, Fairbanks

Rampton, V.N. 1971. Late Quaternary vegetation and climatic history of the Snag-Klutlan area, southwestern Yukon Territory, Canada. *Geological Society of America Bulletin* 82: 959–78

– 1973. The history of thermokarst in the Mackenzie-Beaufort region, Northwest Territories, Canada. *International Association for Quaternary Research, 9th Congress, Abstract Volume*: 299

– 1974. The influence of ground ice and thermokarst upon the geomorphology of the Mackenzie-Beaufort region. In B.D. Fahey and R.D. Thompson, eds, *Proceedings of the 3rd Guelph Symposium on Geomorphology, 1973* (Geoabstracts, Norwich): 43–59

– 1982. *Quaternary Geology of the Yukon Coastal Plain.* Bulletin 317, Geological Survey of Canada, Ottawa

Rampton, V.N., and M. Bouchard. 1975. *Surficial Geology of Tuktoyaktuk, District of Mackenzie.* Geological Survey of Canada, Paper 74–53

Raup, H.M. 1947. Some natural floristic areas in boreal America. *Ecological Monographs*, 17: 221-34

Raup, H.M., and G.W. Argus. 1982. *The Lake Athabasca Sand Dunes of Northern Saskatchewan and Alberta, Canada. 1. The Land and Vegetation.* Publication in Botany 12, National Museums of Canada, Ottawa

Reid, D.E. 1974. *Vegetation of the Mackenzie Valley*, vol. 3. Arctic Gas, Biological Report Series

Richard, P. 1970. Atlas pollinique des arbres et de quelques arbustes indigènes du Québec. III. *Naturaliste Canadien*, 97: 97-161

- 1981. *Paleophytogéographie postglaciaire en Ungave par l'analyse pollinique.* Paleo-Québec 13, Université du Québec à Montréal

Richards, P.W. 1955. *The Tropical Rain Forest.* Cambridge University Press, Cambridge

Ritchie, J.C. 1972. Pollen analysis of Late-Quaternary sediments from the arctic treeline of the Mackenzie River Delta Region, Northwest Territories, Canada. In Y. Vasari, H. Hyvärinen, and S. Hicks, eds, *Climatic Changes in Arctic Areas during the Last Ten Thousand Years* (University of Oulu, Oulu): 253-71

- 1974. Modern pollen assemblages near the arctic tree line, Mackenzie Delta region, Northwest Territories. *Canadian Journal of Botany*, 52: 381-96

- 1976. The late-Quaternary vegetational history of the Western Interior of Canada. *Canadian Journal of Botany*, 54: 1793-1818

- 1977. The modern and late-Quaternary vegetation of the Campbell-Dolomite uplands near Inuvik, N.W.T., Canada. *Ecological Monographs*, 47: 401-23

- 1982. The modern and Late-Quaternary vegetation of the Doll Creek area, North Yukon, Canada. *New Phytologist*, 90: 563-603

Ritchie, J.C., J. Cinq-Mars, and L.C. Cwynar. 1982. L'environnement tardiglaciaire du Yukon septentrional, Canada. *Géographie physique et quaternaire*, 36: 241-50

Ritchie, J.C., and L.C. Cwynar. 1976. Palaeobotany report. In *North Yukon Research Programme, University of Toronto, 1976 Annual Report*

- 1982. The Late-Quaternary vegetation of the North Yukon. In Hopkins 1982: 113-26

Ritchie, J.C., and F.K. Hare. 1971. Late-Quaternary Vegetation and climate near the arctic tree line of northwestern North America. *Quaternary Research*, 1: 331-41

Ritchie, J.C., and S. Lichti-Federovich. 1967. Pollen dispersal phenomena in arctic-subarctic Canada. *Review of Palaeobotany and Palynology*, 3: 255-66

- 1968. Holocene pollen assemblages from the Tiger Hills, Manitoba. *Canadian Journal of Earth Sciences*, 5: 873-80

Ritchie, J. C. and G.A. Yarranton. 1978. The late-Quaternary history of the boreal forest of central Canada. *Journal of Ecology*, 66: 199-212

Rouse, G.E. 1977. Paleogene palynomorph ranges from western and northern Canada. In W.C. Elsik, ed., *Contributions of Stratigraphic Palynology*, vol. 1, *Cenozoic Palynology* (American Association of Stratigraphic Palynologists, Contr. Ser. 5A): 48-65

Rowe, J.S. 1955. *Factors Influencing White Spruce Reproduction in Manitoba and Saskatchewan.* Technical Note 3, Canada Department of Northern Affairs and Natural Resources, Ottawa

- 1966. Phytogeographic zonation: an ecological appreciation. In R.L. Taylor and R.A. Ludwig, eds. *The Evolution of Canada's Flora* (University of Toronto Press, Toronto): 12-27

- 1972. *Forest Regions of Canada.* Department of Environment, Canada Forestry Service, Publication 1300

Rymer, L. 1978. The use of uniformitarianism and analogy in paleoecology, particularly pollen analysis. In D. Walker and J.C. Guppy, eds., *Biology and Quaternary Environments* (Australian Academy of Science): 246–57

Saarnisto, M. 1979. Studies of annually laminated lake sediments. In Berglund 1979a, vol. 2: 61–80

Savile, D.B.O. 1972. *Arctic Adaptations in Plants.* Monograph 6, Canada Department of Agriculture

Schweger, C.S., and T. Habgood. 1976. The late-Pleistocene steppe-tundra of Beringia – a critique. *Abstracts of 4th Conference, American Quaternary Association*, Tempe

Shackleton, J. 1982. *Environmental Histories from Whitefish and Imuruk Lakes, Seward Peninsula, Alaska.* Report 76, Institute of Polar Studies, Ohio State University, Columbus

Shackleton, N.J. and N.D. Opdyke 1973. Oxygen isotope and palaeomagnetic stratigraphy of equatorial Pacific core V28-238; Oxygen isotope temperatures and ice volumes on a 10^5 year and a 10^6 year scale. *Quaternary Research*, 3: 39–55

Sharbatyan, A.A. 1974. *Extreme Estimations in Geothermy and Geocryology* (in Russian). Moscow, U.S.S.R. Academy of Sciences, Institute of Water Problems, Science Publishing House

Shearer, J.M., R.F. MacNab, B.R. Pelletier, and T.B. Smith. 1971. Submarine pingos in the Beaufort Sea. *Science*, 174: 816–18

Shilo, N.A. 1978. The discovery of a mammoth on the Kirgiliakh River, Magadansk region. *Priroda*, 1: 18–20

Shilo, N.A., A.V. Lozhkin, E.E. Titov, and U.V. Shumilov. 1977. A unique new mammoth find. Academy of Sciences, USSR, *Geology*, 236, 4 p.

Simberloff, D. 1980. A succession of paradigms in ecology: essentialism to materialism and probabilism. *Synthèse*, 43: 3–39

Slater, D. 1980. Late Quaternary pollen diagram from the central Mackenzie Corridor area. American Quaternary Association, *Abstracts* volume

Slaughter, C.W., and K.P. Long. 1975. Upland climatic parameters on subarctic slopes in Central Alaska. In G. Weller and S.A. Bowling, eds, *Climate in the Arctic* (University of Alaska, Fairbanks): 276–80

Solomon, A.M., H.R. Delcourt, D.C. West, and T.J. Blasing. 1980. Testing a simulation model for reconstruction of prehistoric forest-stand dynamics. *Quaternary Research*, 14: 275–93

Solonovich, N.G., B.A. Tikhomirov, and V.V. Ykrainstseva. 1977. Preliminary examination of the plant remains in the gut of the Shadrin mammoth (Yakutia). *Trudy Zooloogeskaya Instituta*, 63: 277–80

Sorenson, C.J. 1977. Reconstructed Holocene bioclimates. *Annals of the Association of American Geographers*, 67: 214–22

Spear, R.W. 1980. Late Quaternary history of subalpine and alpine vegetation in the White Mountains of New Hampshire, U.S.A. *Abstracts of Fifth International Palynological Conference, Cambridge*: 371

Spear, R.W. 1983. Paleoecological approaches to a study of treeline fluctuation in the Mackenzie Delta Region, Northwest Territories: preliminary results. In P. Morisset and S. Payette, eds, *Treeline Ecology*, Proceedings of the Northern Quebec Tree-Line Conference, Nordicana, 47, in press

Spetzman, L.A. 1959. *Vegetation of the Arctic Slope of Alaska*. U.S. Geological Survey, Professional Paper 302-B

Stuiver, M. 1970. Long-term C14 variations. In I.V. Olsson, ed., *Radiocarbon Variations and Absolute Chronology, Proceedings Twelfth Nobel Symposium, Uppsala 1969* (Stockholm): 197–213

Sutton, R.F. 1969. *Silvics of White Spruce (Picea glauca (Moench) Voss)*. Forestry Branch Publication 1250, Canada Department of Fisheries and Forestry, Ottawa

Tedrow, J.C.F. 1977. *Soils of the Polar Landscapes*. Rutgers, New Brunswick

Terasmae, J. 1959. Palaeobotanical study of buried peat from the Mackenzie River Delta area, Northwest Territories. *Canadian Journal of Botany*, 37: 715–17

Thaler, G.R., and R.C. Plowright. 1973. An examination of the floristic zone concept with special reference to the northern limit of the Carolinian zone in southern Ontario. *Canadian Journal of Botany*, 51: 1765–89

Thomas, D.C., F.L. Miller, R.M. Russell, and G.R. Parker. 1981. The Bailey Point region and other muskox refugia in the Canadian arctic: a short review. *Arctic*, 34: 34–6

Tikhomirov, B.A. 1958. *The Natural Conditions and Vegetation of the Mammoth Epoch in Northern Siberia* (in Russian). Problems of the North 1, Academy of Science, Moscow

Tolmatchev, A., and B.A. Yurtsev. 1970. Istoriya arkticheskoy flory u ee svyazi s istoriey severnogo ledovitogo okeana. In A. Tolmatchev, ed., *Severnyi ledovityi okean i ego poberezh'e kainozoe* (Leningrad): 87–100

Van Cleve, K., C.T. Dyrness, and L.A. Viereck. 1980. Nutrient cycling in interior Alaska flood plains and its relationship to regeneration and subsequent forest development. In *Forest Regeneration at High Latitudes* (U.S. Department of Agriculture, Forest Service, Report PNW-107): 11–18

Vereschagin, N.K. 1977. The Berelekh 'cemetery' mammoths. In *Mammoth Fauna of the Russian Plain and Eastern Siberia*. (Nauka, Leningrad): 5–50

Viereck, L.A. 1966. Plant succession and soil development on gravel outwash of the Muldrow Glacier, Alaska. *Ecological Monographs*, 36: 181–99

– 1970. Forest succession and soil development adjacent to the Chena River in Interior Alaska. *Arctic and Alpine Research*, 2: 1–26

– 1975. Forest Ecology of the Alaska Taiga. *Proceedings of Circumpolar Conference on Northern Ecology* (National Research Council of Canada, Ottawa): 1–22

Viereck, L.A., and C.T. Dyrness. 1980. *A Preliminary Classification System for the Vegetation of Alaska*. General Technical Report PNW-106, U.S. Department of Agriculture, Forest Service

Viereck, L.A. and E.L. Little, Jr. 1972. *Alaska Trees and Shrubs*. U.S. Forest Service, Agriculture Handbook, 410

– 1975. *Atlas of United States Trees*, vol. 2. *Alaska Trees and Common Shrubs*. U.S. Forest Service, Misc. Publ. 1293

Viereck, L.A., and L.A. Schandelmeier. 1980. *Effects of Fire in Alaska and Adjacent Canada – A Literature Review*. U.S. Department of the Interior, BLM-Alaska, Technical Report 6

Von Rudolph, E., E.T. Oswald, and E. Nyland. 1981. Chemosystematic studies in the genus *Picea*. V. Leaf oil terpene composition of white spruce from the Yukon Territory. *Canadian Forestry Service Research, Notes* 1: 32–4

Waldron, R.M. 1965. Cone production and seedfall in a mature white spruce stand. *Forestry Chronicle*, 41: 314–29

Walker, D. 1982a. Vegetation's fourth dimension. *New Phytologist*, 90: 419–29

Walker, D. 1982b. The development of resilience in burned vegetation. In E.I. Newman, ed., *The Plant Community as a Working Mechanism* (Special Publication 1, British Ecological Society): 27–43

Washburn, A.L. 1973. *Periglacial processes and Environments.* Edward Arnold, London

Watson, E. 1971. Remains of pingos in Wales and the Isle of Man. *Geological Journal*, 7: 381–92

Watts, W.A. 1973. Rates of change and stability in vegetation in the perspective of long periods of time. In H.J.B. Birks and R.G. West, eds, *Quaternary Plant Ecology* (Blackwells, Oxford): 195–206

Webb, T. III. 1981. The past 11,000 years of vegetation change in eastern North America. *Bioscience*, 31: 501–6

Webb, T. III., and D.R. Clark. 1977. Calibrating micropaleontological data in climatic terms: a critical review. *Annals New York Academy of Science*, 288: 93–118

Webber, P.J. 1974. Tundra primary productivity. In J.D. Ives and R.G. Barry, eds., *Arctic and Alpine Environments* (Methuen, London): 445–73

Webber, P.J., and D.A. Walker. 1975. Vegetation and landscape analysis at Prudhoe Bay, Alaska: A vegetation map of the Tundra Biome study area. In J. Brown, ed., *Ecological Investigations of the Tundra Biome in the Prudhoe Bay Region, Alaska* (Biological Paper, University of Alaska, Special Report 2): 81–91

Welsh, S.L., and J.K. Rigby. 1971. Botanical and physiographic reconnaissance of northern Yukon. *Brigham Young University, Science Bulletin, Biological Series* 14: 1–16

Wiggins, I.L. 1951. The distribution of vascular plants on polygonal ground near Point Barrow, Alaska. *Contributions of the Dudley Herbarium*, 5: 69–95

Wiggins, I.L., and J.H. Thomas. 1962. *A Flora of the Alaskan Arctic Slope.* Arctic Institute of North America, Special Publication 4, University of Toronto Press, Toronto

Wolfe, J.A. 1966. *Tertiary Plants from the Cook Inlet Region, Alaska.* U.S. Geological Survey, Professional Papers B32, 398

– 1969. Neogene floristic and vegetation history of the Pacific Northwest. *Madrono*, 20: 83–110

– 1971. Tertiary climatic fluctuations and methods of analysis of Tertiary floras. *Palaeogeography, Palaeoclimatology, Palaeoecology*, 9: 27–57

– 1972. An interpretation of Alaskan Tertiary Floras, A. Graham, ed., *Floristics and Paleofloristics of Asia and Eastern North America* (Elsevier, Amsterdam): 201–33

– 1975. Some aspects of plant geography of the Northern Hemisphere during Late Cretaceous and Tertiary. *Annals Missouri Botanical Garden*, 62: 264–79

– 1977. *Paleogene Floras from the Gulf of Alaska Region.* U.S. Geological Survey, Professional Paper 997

– 1981. A chronological framework for Cenozoic megafossil floras of northwestern North America and its relation to marine geochronology. *Geological Society of America, Special Paper* 184: 39–4

Wolfe, J.A., D.M. Hopkins, and E.B. Leopold. 1966. *Tertiary Stratigraphy and Paleobotany of the Cook Inlet Region, Alaska.* U.S. Geological Survey, Professional Papers 398-A, A1 to A29

Wolfe, J.A., and E.B. Leopold. 1967. Neogene and Early Quaternary vegetation of northwestern North America and northeastern Asia. In D.M. Hopkins, ed., *The Bering Land Bridge* (Stanford, University Press, Stanford): 193–206

Wright, H.E. Jr. 1968. A square-rod piston sampler for lake sediments. *Journal of Sedimentary Petrology*, 37: 975–6

– 1980. Cores of soft lake sediments. *Boreas*, 9: 107–14

Wright, J.W. 1964. Flowering age of clonal and seedling trees as a factor in choice of breeding system. *Silvea Genetica*, 13: 21–7

Yarranton, G.A. and J.C. Ritchie 1972. Sequential correlations as an aid in placing pollen zone boundaries. *Pollen et Spores*, 14: 253–71

Yorath, C.J., H.R. Balkwill, and R.W. Klassen. 1969. Geology of the eastern part of the northern interior and arctic coastal plains, Northwest Territories. Geological Survey of Canada, Paper 68-27: 1–26

Young, S.B. 1971. *The Vascular Flora of St. Lawrence Island with Special Reference to Floristic Zonation in the Arctic Regions.* Contributions Gray Herbarium, Harvard University, 201

– 1976. Is steppe tundra alive and well in Alaska? *American Quaternary Association, Abstracts*, 4: 84–8

Yurtsev, B.A. 1963. On the floristic relations between steppes and prairies. *Botaniska Notiser*, 116: 396–408

– 1972. Phytogeography of Northeastern Asia and the problem of trans-Beringian floristic interrelations. In A. Graham, ed., *Floristics and Paleofloristics of Asia and Eastern North America* (Elsevier, Amsterdam): 19–54

– 1973. The phytogeographic zonation and floristic regions of Chukotka. *Botanical Journal*, 58: 945–64

– 1974. *Phytogeographical Problems in Northeastern Asia.* Nauka, Leningrad

– 1978. The botanical-geographical character of southern Chukatki. *Komarov Readings*, vol. 26 (Vladivostok, Academy of Sciences, USSR): 3–62

– 1981. *The Relict Steppe Complexes of Northeastern Asia* (in Russian). Siberian Publishing House of 'Science,' Novosibirsk

Yurtsev, B.A., A.I. Tolmatchev, and O.V. Rebristaya 1978. The floristic delimitation and subdivision of the arctic. In B.A. Yurtsev, ed., *Arkicheskaya floristicheskaya oblast* (Academy of Science, Leningrad): 9–104

Zasada, J.C. 1980. Some considerations in the natural regeneration of white spruce in Interior Alaska. In *Forest Regeneration at High Latitudes* (General Technical Report PNW-107, U.S. Department of Agriculture, Forest Service): 25–9

Zasada, J.C., and R.A. Gregory. 1969. *Regeneration of White Spruce with Reference to Interior Alaska: A Literature Review.* Research Paper PNW-70, 37, U.S. Department of Agriculture, Forest Service, Juneau, Alaska

Zasada, J.C., and D. Lovig. 1983. Observations on primary dispersal of white spruce seed. *Bulletin of the Pacific Northwest Forest and Range Experiment Station*, in press

Zoltai, S.C. 1973. The range of Tamarack (*Larix laricina* (DuRoi) K. Koch) in Northern Yukon Territory. *Canadian Journal of Forest Research*, 3: 461–4

Zoltai, S.C., D.J. Karasiuk, and G.W. Scotter. 1980. *A Natural Resource Survey of the Thomsen River Area, Banks Island, Northwest Territories.* Parks Canada

Zoltai, S.C., and W.W. Pettapiece. 1973. *Terrain, Vegetation and Permafrost Relationships in the Northern Part of the Mackenzie Valley and Northern Yukon.* Environmental-Social Committee Northern Pipelines, Task Force of Northern Oil Development, Report no. 73–74

– 1974. Tree distribution on perennially frozen earth hummocks. *Arctic and Alpine Research*, 6: 403–11

Zoltai, S.C., and C. Tarnocai. 1974. *Soils and Vegetation of Hummocky Terrain.* Environmental-Social Committee Northern Pipelines, Task Force on Northern Oil Development, Report no. 74–5

– 1975. Perennially frozen peatlands in the Western Arctic and Subarctic of Canada. *Canadian Journal of Earth Sciences*, 12(1): 28–43

Zoltai, S.C., and H. Zalasky. 1979. Postglacial fossil tamarack (*Larix laricina*) wood from the Mackenzie Delta, N.W.T. *Forestry Service Bi-monthly Research Notes*, 35: 7–8

Index

An index to topics, to the more important plant names in the text, and to the authors of contributions of direct significance to the main themes

www.ingramcontent.com/pod-product-compliance
Lightning Source LLC
Chambersburg PA
CBHW030459210326
41597CB00013B/726